The Alfred C. Glassell, Jr.—
University of Miami

ARGOSY EXPEDITION
to ECUADOR

Part 1: Introduction and Narrative

STUDIES IN TROPICAL OCEANOGRAPHY

No. 1. Systematics and Life History of the
Great Barracuda, *Sphyraena barracuda* (Walbaum)
By Donald P. de Sylva

No. 2. Distribution and Relative Abundance of
Billfishes (*Istiophoridae*) of the Pacific Ocean
By John K. Howard and Shoji Ueyanagi

No. 3. Index to the Genera, Subgenera, and Sections of
the Pyrrhophyta
By Alfred R. Loeblich, Jr. and Alfred R. Loeblich, III

No. 4. The R/V *Pillsbury* Deep-Sea Biological Expedition to
the Gulf of Guinea, 1964–1965 (Parts 1 and 2)

No. 5. Proceedings of the International Conference on
Tropical Oceanography, November 17–24, 1965, Miami
Beach, Florida

No. 6. American Opisthobranch Mollusks
By Eveline Marcus and Ernst Marcus

No. 7. The Systematics of Sympatric Species in West Indian
Spatangoids: A Revision of the Genera *Brissopsis*,
Plethotaenia, *Paleopneustes*, and *Saviniaster*
By Richard H. Chesher

No. 8. Stomatopod Crustacea of the Western Atlantic
By Raymond B. Manning

No. 9. Effects of Abatement of Domestic Sewage
Pollution on the Benthos, Volumes of
Zooplankton, and the Fouling Organisms
of Biscayne Bay, Florida
By J. Kneeland McNulty

No. 10. Investigations on the Gray Snapper,
Lutjanus griseus
By Walter A. Starck, II and Robert E.
Schroeder

Argosy under sail (Victor Shifreen).

Studies in Tropical Oceanography No. 11 (Part 1)
Rosenstiel School of Marine and Atmospheric Science
University of Miami

The Alfred C. Glassell, Jr.—
University of Miami

ARGOSY EXPEDITION to ECUADOR

Part 1: Introduction and Narrative

by DONALD P. DE SYLVA

UNIVERSITY OF MIAMI PRESS

Coral Gables, Florida

This volume may be referred to as:
Stud. trop. Oceanogr. Miami 11
(Part 1): 164 pp. 91 figs. July,
1972

Contribution 1211, University of Miami,
Rosenstiel School of Marine and Atmospheric Science

Contents

Figures   *9*

Acknowledgments   *13*

Introduction   *17*

Material and Methods   *21*

Narrative of the Cruise   *23*

Oceanographic Observations   *109*

Appendix (Tables 1, 2, and 3)   *123*

Bibliography   *133*

Figures

Argosy under sail *frontispiece*

1 Cruise track of *Argosy* *24*
2 Transferring equipment off Panama *26*
3 Shrimp boats in Panama Bay *27*
4 Islet in Santelmo Bay *28*
5 Night-lighting in Santelmo Bay *30*
6 Surge channels in north cove of Bahía Piñas *31*
7 *Argosy* at anchor, Bahía Piñas *32*
8 Preparing to set pyramidal trap and gill net in Bahía Piñas *33*
9 Mouth of Río Jaqué off Punta Jaqué *34*
10 Village of Jaqué, Panama *35*
11 Seining off Jaqué *35*
12 Location of *Argosy* collections near Jaqué *36*
13 Twenty-meter seine *36*
14 Preparing to dive off Bahía Piñas *38*
15 Location of *Argosy* Collection No. 15 *39*
16 Moorish idol *40*
17 Trawl winch operated from *Argosy* *41*
18 Mako shark *42*
19 *Sea Quest* returns with shark *43*
20 Towing dead sharks *44*
21 Sorting a trawl catch *45*
22 Captain Arved J. Rosing *46*

23   Yellowfin tuna   *47*
24   Silky shark   *48*
25   Young dolphin   *49*
26   Tony, with friend   *50*
27   Young sailfish   *51*
28   Yellow-banded sea snake   *51*
29   Location of *Argosy* collections at Bahía Solano, Colombia   *53*
30   Snapper caught by Captain Red Stuart   *54*
31   Plankton tow from off Colombia   *55*
32   Preparing specimens   *56*
33   Marketplace at Buenaventura   *57*
34   "Gonzalez"   *58*
35   A moment's rest   *58*
36   Shrimp fleet in Bahía de Buenaventura   *59*
37   Hoisting sails off Bahía de Buenaventura   *60*
38   Grunts and goatfish at Gorgona Island   *61*
39   Snappers and grouper at Gorgona Island   *62*
40   A rare cirrhitid   *62*
41   Mr. Glassell and sailfish   *63*
42   Preparing to dive off Gorgona Island   *64*
43   Fighting a striped marlin   *65*
44   Red Hagen bills a sailfish   *66*
45   A 325-lb. black marlin is examined   *67*
46   Flounder   *69*
47   From the harbor of Tumaco, Colombia   *70*
48   Sightseeing in Tumaco   *71*
49   In the harbor of Manta, Ecuador   *74*
50   Northwest coast of Isla de la Plata   *75*
51   Looking south along eastern side of La Plata   *76*
52   Looking north from *Argosy*'s anchorage   *77*
53   Crew of bongo boat   *77*
54   Hosts at La Plata   *78*
55   Tidepool collecting at La Plata   *78*
56   Dr. F. G. Walton Smith   *79*
57   Señor Pedro Lucas, Mr. Glassell, and Dr. Smith   *80*

58 Looking westward from off La Plata *81*
59 Bluefaced booby *81*
60 Mr. Glassell with sailfish and striped marlin *82*
61 Vic Shifreen with a broomtail grouper *83*
62 Tuna clippers in Manta Harbor *87*
63 Ashore for supplies *87*
64 Water casks and burros in Manta *88*
65 Isla de la Plata, Ecuador *90*
66 Bathythermograph off La Plata *92*
67 Parrotfish *93*
68 Mako shark *94*
69 Mako shark *95*
70 Small tunas make bait for mako *96*
71 Garden eel *97*
72 Congrid eel *98*
73 Mr. Glassell and Red Hagen display roosterfish *99*
74 Catch off Salango Island *100*
75 Cornetfish *101*
76 Snapper *102*
77 Dr. Smith and Mr. Glassell atop La Plata *105*
78 Scientific party and crew at end of expedition *107*
79 Location of oceanographic transects off Ecuador *111*
80 Depth contours in collecting areas off coastal Ecuador *112*
81 Bathythermograph traces off Ecuador *113*
82 Thermocline depth off Ecuador *114*
83 Temperature structure at section A-A$_1$ *115*
84 Temperature structure at section B-B$_1$ *116*
85 Temperature structure at section C-C$_1$ *116*
86 Surface water temperature off Ecuador *117*
87 Temperature at depth of 10 m off Ecuador *118*
88 Temperature at depth of 20 m off Ecuador *119*
89 Temperature at depth of 30 m off Ecuador *120*
90 Salinity structure at section B-B$_1$ off Ecuador *121*
91 Salinity structure at section C-C$_1$ off Ecuador *121*

Acknowledgments

Many people aided the expedition. Among these, we wish to thank Mr. Manuel J. Castillo, consul general of Panama in Miami, and the officials of the government of the Republic of Panama for arranging for and permitting us to collect in their waters. Similarly, Mr. Mario Iragorri, consul general of Colombia, and Mr. Hugo Nichols, vice-consul, gave us invaluable assistance in procuring necessary papers and travel permission.

Mr. José R. Chiriboga, formerly consul general of Ecuador in Miami, was particularly helpful in arranging permission for us to enter the coastal waters of Ecuador and to make collections there.

We wish to thank especially the officials of these countries for sanctioning the expedition, and we trust that the information we have obtained will be of use to them in the development and utilization of their rich marine resources.

The Alfred C. Glassell, Jr.—University of Miami

ARGOSY EXPEDITION to ECUADOR

Part 1: Introduction and Narrative

Introduction

The eastern tropical Pacific is extremely interesting to the marine biologist and ecologist because of its faunal similarity to the western tropical Atlantic and its separation from the western and central tropical Pacific by vast expanses of deep water. Systematists and evolutionists require comparative material to study sibling species, rates of evolution, and zoogeographical affinities of this fauna with other areas. Ekman (1953:33 ff.) reviewed in detail the processes and results of isolation due to the formation of the Panamanian isthmus during several geologic periods as recently as the late Pliocene or early Pleistocene and the implications of the eastern Pacific barrier. Briggs (1961) discussed the effect of the land barrier of the western hemisphere on the distribution of shore fishes.

Since Ekman's classic work on the zoogeography of the eastern tropical Pacific, outstanding collections have been reported in a continuous series of publications by the Allan Hancock Foundation, the New York Zoological Society, the Academy of Natural Sciences of Philadelphia, the Vanderbilt Museum, and many others. I have included papers emanating from these expeditions in the bibliography.

Recently, a proposal to dredge a sea-level canal across the Panamanian isthmus using nuclear energy has raised the questions of subsequent faunal and floral interchange between the Atlantic and Pacific oceans and its consequences, and, especially, the possible direct and indirect effects of atomic radiation upon the ecology, biology, and genetic makeup of the organisms in this area. Voss (1967) summarized the problems inherent in the proposed sea-level canal and documented the importance of the area both as a source of food and for scientific study.

The regions emphasized historically in previous reports include the Pacific coast of Mexico and Baja California, including the Gulf of California, Pacific Panamanian waters, and the Clipperton, Cocos, Revillagigedo, and Galápagos islands. Less studied are the littoral and shallow waters between the Panama border to the Gulf of Guayaquil, Ecuador. Here, vast stretches of jungle and impenetrable coast or steep foothills encroaching to the water's edge have made collecting from shore difficult and impractical. The small islands off this coastal area, including the Islas de las Perlas (Panama), Isla de Gorgona (Colombia), and Isla de la Plata (Ecuador), are of interest because of the extensive habitat that they offer for coastal forms, and the shelter on their lee sides that affords luxuriant coral formations. These formations provide opportunities for scientific collection that generally cannot be undertaken on the mainland. Here, the exposure to the prevailing southwest swell, and the heavy surge that results make shallow-water collecting difficult and dangerous.

These coastal islands harbor a fauna that is essentially eastern tropical Pacific, but the lack of intensive collections in these areas south of Panama with various types of collecting gears precludes as complete a knowledge of the fauna as might be hoped for. New or rare species and range extensions from the north or even from the western and central tropical Pacific are more likely to be found in such an area as intensive collecting is undertaken.

Because of its close relation to the western tropical Atlantic, the fauna of the eastern tropical Pacific is of interest to the Institute of Marine Sciences (now the Rosenstiel School of Marine and Atmospheric Science) of the University of Miami. Students of several groups of fishes and invertebrates have not been able to comprehend fully the phylogeny and relationships of some species nor the limits within genera and even families, and the variation throughout the range of taxa has been poorly studied because adequate material from other areas has not been available. Secondly, a number of ecological problems associated with the systematics of these groups are in need of study and comparison with other tropical areas. Finally, the life histories of a number of marine forms might be better understood if adequate data from other areas were available to provide comparisons from slightly different ecological habitats.

Here, then, was an area that had not been studied intensively. Although areas to the north had long been investigated, collections had been made years ago and modern collecting equipment and methods were consequently not available to the scientists. Most specimens extant in museums have been taken by seines, trawls, and dredges, while larger fishes were generally purchased in markets. The recent development of self-contained diving gear (scuba) that permits collecting in deeper water and the use of improved poisons such as rotenone to collect fishes and invertebrates offer an efficient and nonselective method to collect burrowing and coralligerous species not obtainable by conventional gear and methods. While many shallow-water tide pool species, which have predominated in the literature, occur only to rather shallow depths, collecting below depths of even 15 m may yield new or rare species. Thus the institute wished to collect intensively in several localities between Panama and Ecuador using many types of gears in different habitats and at different times of the day.

For many years Mr. Alfred C. Glassell, Jr., of Houston, Texas, had expressed a long-standing desire to sponsor a collecting expedition to this area. In 1960 plans were made between Mr. Glassell and the Institute of Marine Sciences to undertake a six-week expedition to the waters of Panama, Colombia, and Ecuador. Due to the generosity of Mr. Glassell, it was possible to outfit a team of scientists who were trained scuba divers with adequate equipment and facilities and to supply them with several kinds of nets, dredges, and trawls to obtain specimens for study. Mr. Glassell has also made possible the analysis and publication of the results of the expedition. The material obtained amounts to a bulk lot of over twelve hundred liters of fishes and invertebrates.

The auxiliary ketch *Argosy* (Frontispiece) was made available to the University by Mr. Glassell. She is a 32.6-m steel motor sailer with twin diesels, and was outfitted with a small winch for plankton towing, trawling, and bathythermograph casts; it was also used as a home base for operations. A second vessel, the *Sea Quest*, a 12-m twin-screw sportfisherman, was chartered from and captained by the late Captain Stirling R. ("Red") Stuart of Miami Beach. She was utilized mainly for obtaining the larger pelagic fishes which could not be taken by conventional collecting gear.

During our stay in Panama we were ably assisted by Señores Gabriel Gamboa, Osvaldo Sánchez, and Roberto Ruíz, of the Department of Fisheries, and we wish to thank them for their aid.

Mr. William Saenz, Bogotá, Colombia, was instrumental in making numerous arrangements for our crew and vessels to enter Colombian waters. During our stay in Buenaventura and during subsequent field operations, the assistance rendered by Mr. Saenz and by Dr. Luis Ortíz Borda and Mr. Eric Saenz was invaluable.

Thanks are due to Dr. F. Bourgois, representative of the Food and Agriculture Organization of the United Nations in Guayaquil, Ecuador, and Director, Instituto National de Pesca, for his great assistance in arranging entrance into the coastal waters for collecting, and to Mr. Robert W. Ellis, Food and Agriculture Organization, for his assistance in a number of ways. At Manta, Mr. Robert Carpenter of INEPACA supplied us with valuable information on local conditions, and Mr. Carlos Carrera of Guayaquil was of considerable help to us in many ways. At Isla de la Plata, Sr. Guillermo Lucas and the late Sr. Emilio Estrada permitted us to use their facilities and made our stay at La Plata a profitable one.

Arrangements for food, fuel, and supplies for the expedition were facilitated through Boyd Brothers Steamship Agency, Ltd., Cristobal; by Grace y Cía., Buenaventura, Colombia; and by the Anglo-Ecuadorian Oilfields, Ltd., in La Libertad, Ecuador.

Throughout this expedition we were personally assisted by Mr. Glassell and his associate Mr. Victor Shifreen of Houston, Texas. The late Captain Arved J. Rosing and his competent crew of the yacht *Argosy* made the trip a safe and profitable one, and they personally assisted us in collecting data and specimens. Similarly, the late Captain Stirling R. Stuart, captain of the vessel *Sea Quest*, and his crew supplied us with numerous specimens.

Prior to the expedition, Mr. and Mrs. John A. Manning, Coral Gables, Florida, supplied us with valuable information about water conditions in the vicinity of Isla de la Plata.

Finally, thanks are due to our colleagues at the Institute of Marine Sciences for expediting plans for the expedition, particularly to William Barkley, former port captain at the Institute of Marine Sciences, for expediting construction of special gear and shipping of the vessel *Sea Quest*, and to Mary Manning Morford for considerable assistance in the purchasing of equipment. Drs. Anthony J. Provenzano, Jr., and Saul Broida offered valuable comments on the manuscript.

Material and Methods

The equipment used in the field was generally of standard design. Collections during scuba operations and in tidepools were made using 5% emulsifiable rotenone sold under the trade name "Pronoxfish," and it was used as sparingly as possible. Generally two liters, in plastic squeeze bottles, were needed to sample a reef area during collecting operations, and the samples were taken at mid-ebb tide whenever possible. Tide pool collections averaged about one liter.

Trawls used were of two kinds: a balloon-type trawl, measuring 5 m across along the head and foot ropes, with 2-cm bar mesh (cotton) and a 3-mm nylon liner sewn in the last part of the cod end; and a flat trawl of identical dimensions and of the same mesh.

The gill net was a 270-m nylon multimeshed net, consisting of six 45-m sections, each of a different sized mesh, hung together. The mesh sizes were graded from 2 to 10 cm.

The plankton nets used were 1-m nylon mesh, size 00, made by Mr. W. C. Schroeder of Falmouth, Massachusetts. The net was towed at the surface either abeam or aft of *Argosy*, and most collections were made at night. In all cases, the net was towed with the rim about 10 cm out of water.

Other plankton collections were made using a modified Gulf-I high-speed plankton sampler (HSPS) described by Smith, de Sylva, and Livellara (1964). The intake of the sampler is 5 cm, and the sampler was towed at an average speed of 8 knots, the cruising speed of *Argosy*.

Night-light collections were made by suspending a 150-watt light bulb encased in a waterproof housing about 1/2 m underwater. In some cases, a rheostat was connected to the light source to vary the intensity, and as the fish schools became more concentrated, they could be dipnetted more easily. No quantitative data on this aspect were maintained. Collections were made using dip nets with 6-mm knit cotton netting, or 2-mm plastic screen.

Beach seining was done with a nylon bag seine measuring 18 m long by 2 m deep, with a 6-mm mesh in the wings and 3-mm mesh in the bag.

A pyramidal trap net measuring 1.2 m on each side, designed by the Scripps Institution of Oceanography, was used with some success for fishes and invertebrates. Our traps were modified somewhat and were not free-ascending; they were attached to the winch cable of *Argosy* or left in rocky or coral areas.

Hydrographic data were taken with a 270-m bathythermograph calibrated against a surface bucket thermometer. Salinity samples were taken from deep strata with Nansen bottles and the results titrated in the laboratory in Miami.

It is beyond the ability of the staff of the school to try to report on all groups of organisms which we collected. Collections have been sorted to major groups (i.e., families, orders) and shipped to specialists for analysis. Each specialist is publishing his findings in one or more papers as part of a continuing series on the expedition to be published at the School of Marine and Atmospheric Science, University of Miami. As each paper is received, it will be reviewed and published in the format of the expedition report, and the final results will be bound in one or more volumes as a contribution to the series. *Studies in Tropical Oceanography*. The writer served as scientist-in-charge of the expedition after its inception in June 1961, and is responsible for the accuracy of the overall account of the trip.

Type specimens will be deposited with the U.S. National Museum and the Academy of Natural Sciences of Philadelphia unless noted otherwise.

A list of collecting stations, previously reported by de Sylva (1963), is appended as Table 1 (p. 124). The weather log of *Argosy* during the expedition is also included as Table 2 (p. 130).

Narrative of the Cruise

5 SEPTEMBER 1961

The scientific party, Mr. Glassell, and Mr. Victor Shifreen, our photographer, departed at about 1300 hours from the Miami International Airport. Several weeks earlier in Miami, we had loaded *Argosy* and *Sea Quest* with collecting gear, jars, equipment of all sorts, and as much food as we could stow. The airplane flight was a rough one, and although we were flying at a height of nearly 9 km, the powerful vortex of Hurricane Carla was churning below us across the Gulf of Mexico, and we encountered turbulent weather and dense cloud cover for the entire flight. We arrived at Tocumen Airport, Panama, claimed our baggage, and headed for our hotel and the tallest rum punch we could find.

Shortly after our arrival, we contacted Señores Gabriel Gamboa, Osvaldo Sánchez, and Roberto Ruíz. Señor Gamboa is a graduate of the Instituto de Ciencias Policiales de la República Mexicana and is now with the Laboratorio Pesca of the Republic of Panama. These men planned to accompany us on part of the cruise in hopes that they would be able to see a variety of gear and collecting methods in action. In turn, we welcomed their knowledge of the marine resources of Panama. Plans were made that afternoon for their trip and for ship clearance.

6 SEPTEMBER 1961

The day was spent making last-minute arrangements, purchasing additional ship supplies, buying souvenirs in Panama City, and purchasing study specimens in the fish market. Most specimens we found in the market were Pacific species, but the merchants' insistence that some individuals, known scientifically only from the Atlantic, were from the Pacific, led us to question

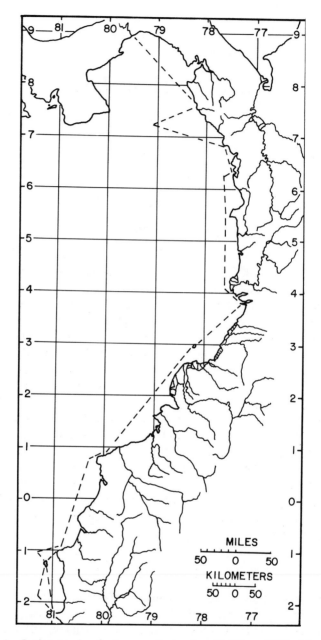

Figure 1. Cruise track of *Argosy*, September–October, 1961 (drawing by Richard Marra).

the source, and we decided that a few were brought over from Colon. We obtained some fresh specimens of snook (*Centropomus* spp.), barracudas (*Sphyraena* spp.), flounders, trouts, and croakers (*Cynoscion, Odontoscion, Menticirrhus*), greenbar jacks (*Caranx caballus*), blue runners (*Caranx crysos*), sierras (*Scomberomorus sierra*), black skipjack (*Euthynnus lineatus*), and threadfin (*Polydactylus*). Apparently all of these are choice food items.

7 SEPTEMBER 1961

There were dense gray clouds in all quarters of the sky as we departed from Balboa, Canal Zone, aboard *Argosy* for parts unknown. The cruise track to be followed was largely coastal (Fig. 1) but some offshore work was planned. Our ultimate goal was Isla de la Plata, off the northwestern coast of Ecuador, where we hoped to find many types of game fish and reef fish. Large marlin had been seen, and Mr. Glassell expected to break his own world record for a 1560-lb (710-kg) black marlin which he had set eight years earlier off Cabo Blanco, Peru. We were eager to collect in an area where relatively little marine research had been carried out, and we were excited about diving in waters— from Panama to Ecuador—where few men had previously ventured. We were especially concerned, however, because a month earlier a tour boat from Buenaventura, Colombia, had keeled over and most of its passengers had drowned or, more frightening to us, been devoured by the legendary packs of sharks which roam the shore of the eastern tropical Pacific.

Argosy slid out through the harbor at Balboa carrying seventeen of us, with Red Stuart's 12-m *Sea Quest* alongside (Fig. 2). To the six scientists aboard, this expedition was a real challenge. Dr. Walter A. Starck, Jr., then a graduate student at the Miami institute, was well known for his extensive underwater experience. His goal was to collect reef fishes and to study their ecology. Dr. Anthony J. Provenzano, Jr., then also a graduate student at Miami, was primarily concerned with crustaceans, particularly hermit crabs and their systematics and development. And his close friend, Dr. Clyde A. Roper, was studying invertebrates, especially squids, and working on his graduate degree at Miami. Dr. Dennis Paulson, now at the University of Washington, was observing and collecting birds, snakes, lizards, mammals, frogs, insects, fishes, and whatever else he could add to the knowledge of science. I was observing and collecting the large pelagic fishes taken on the expedition, larval fishes obtained from plankton tows, and fishes collected about the reefs and estuaries. A longtime associate of Mr. Glassell, Victor Shifreen, was in charge of photography. Mr. Glassell obtained specimens and

Figure 2. Transferring equipment between stations off Panama (Vic Shifreen).

data for us and assisted us in many other ways during the six-week period. The remainder of the group consisted of the master of the *Argosy*, Captain Rosing, and his crew of five, and Captain Stuart and his crew of two aboard *Sea Quest*. Also with us were the three Panamanian fisheries officers, Señores Gamboa, Sánchez, and Ruíz.

The Panamanian coastline is a rugged one, yet lush and rich. Verdant islands line the harbor off Balboa (Fig. 3) where the extensive Panama shrimp fleet lies at anchor. We learned that the major fishery here is for the *camaron rojo, Penaeus brevirostris*, which is caught at about 70 m. Also taken, but less desirable, are *Sicyonia* sp., *Solenocera* sp., and *Trachypenaeus similis*.

Our course was 138° as we left Balboa harbor at 0600 hours, changing soon to 148°. Heavy fractocumulus clouds seemed to be everywhere, but no squall had yet occurred. At 0720, we lowered the modified Gulf I-A high-speed plankton sampler (HSPS) from the stern and towed it approximately 20 m behind the boat and about 2 m below the surface. With its long, steady southwest swell, the sea was at first smooth and inviting. Wilson's petrels, laughing gulls, frigate birds, and brown boobies seemed to be everywhere. But the sea was now becoming rougher, blacker, and eerier.

At 0850, we brought in our first collection of the six-week expedition, and we were delighted to find that we had obtained a fine selection of plankton, including a larva which subsequently proved to be one of the tunas (*Thunnus* sp.), about 2 cm long. Young mullet, copepods, zoeal stages of shrimp, stomatopods, and siphonophores were common in the jar. But the tow was poor because of the 5-cm opening of the sampler's mouth, and after a 1½-hour tow, less than 1 cm of plankton had settled at the bottom of the one-liter jar, indeed a lean catch.

From 0900 to 1100 little life was seen, but shortly after 1100 five large porpoises (*Tursiops* sp.) began to follow *Argosy*. No fishes or birds were observed in the area until 1120, when a school of large flyingfish soared by.

At 1300, Mr. Glassell transferred from *Argosy* to *Sea Quest* to speed on ahead to troll for billfish. Angling took place about 15 km from Isla Rey, in the Pearl Islands (Islas Perlas) group. A 42-kg Pacific sailfish (*Istiophorus greyi*) was immediately taken by him, and several dolphin (*Coryphaena hippurus*) were seen.

Shortly after 1400 hours, we began to round the tip of Isla San José, one of the beautiful Islas Perlas. These are a series of small, volcanic island outcroppings, and most are low, perhaps 100 m high, with steep scarp faces on the seaward side (Fig. 4). Long, low rollers, today pushed by a force-3 wind, batter these beaches continuously, making it very difficult for shore collecting of any type. Visibility is poor now, and there are no signs of fish or bird life.

At 1600, we arrived in Islas Perlas and anchored in 10 m in Santelmo Bay at the volcano-covered Isla del Rey. By now, the wind had diminished and the

Figure 3. Shrimp boats in Panama Bay (Donald de Sylva).

Figure 4. Islet in Santelmo Bay at Isla del Rey, Islas Perlas, Panama (Walter Starck).

sea had become smoother, but always there is that gentle but persistent southwest swell so characteristic of this part of the world. The greenery and vines hanging from the steep cliffs reminds one of Robinson Crusoe's mysterious haunts. Tropical hanging vines intertwine with the *Acacia*-like trees at the tip of the bluffs, and long-bladed sawgrass droops along the bedded limestone bluff; all are desperately staving off the eventual erosion of the beach by the encroaching surf and sand. Despite the greenery, there is little drinking water, and the twenty-five or so houses that comprise the fishing village of Mafafa, apparently the only village, must be supplied with water by barges from the mainland. Santelmo Bay is crescent shaped, with coconut palms scattered above the strandline. The beach itself is steep, perhaps with a slope of 15° to 20°, and the volcanic sand is coarse and dark brown. Surf is strong, but a few rock-sheltered tide pools offer some shelter from the heavy combers.

By 1630, we were ashore in the little inboard lugger, but the surf prevented us from beaching the boat, and gear had to be swum in or passed by bucket brigade. A collection was made with the 20-m fine-mesh seine along the crescent beach. The low stage of the tide afforded us an easier time, but the surf, undertow, and sharp beach drop-off made it difficult for us to collect the specimens. A sand diver (*Dactyloscopus*), several small pompano (*Trachinotus*), and thousands of postlarval anchovies (*Cetengraulis?*) were taken in the surf. Only a few anchovies could be preserved, and the remainder were immediately snapped up by what appeared to be carangids which moved into feed and by birds which suddenly appeared.

Our next station (*Argosy* Collection No. 3), begun at about 1630, was in the tide pool and surge channels at the east end of the beach. Clyde and Walter, using snorkels, spread the poison on the bottom of the pools, but they were quite badly dashed about by the surf. Within minutes, the small secretive fishes, mostly gobies and blennies, began reacting to the poison and popping out of the water. They darted back and forth from one tide pool to the next in search of untainted waters. Soon, morays, grunts, mullet, silversides, and pomacentrids similar to *Pomacentrus leucostictus* floated to the surface where they were quickly dipnetted by us. Then Starck became frightened by what he thought was a large shark in one of the surge channels and resumed his collection cautiously. By 1715, a few gobies were still darting about the tide pools when suddenly many clingfish flipped about as the tide ebbed, and these were promptly pickled.

Two native fishermen in a *canoa* loaded with rice arrived to watch. They were kind enough to tow the deep end of our seine offshore and back onto the beach, and we gave them the few edible fish we had: the larger grunts, jacks, and pompanos; we kept the smaller specimens. They then became very interested in watching us collect, and promptly walked along the tide pools in search of specimens which they saved for us, including holothurians, an *Arbacia*-type urchin, several of the many large chitons (*Fissurella*), *Balanus* (*amphitrite*), and some *Littorina*-like gastropods.

The talus slope below the volcanic caves comprised a rich variety of rock formations. Breccias, sills, conglomerates, cupric ore, magma of all types, lava flows, and sandstone all provided a most decorative setting behind the sand beach. Dennis Paulson walked along the beach to collect lizards and crabs with his BB gun and had found a number of iguana and turtle tracks along shore, as well as a tidal estuary and mangrove swamp that was full of mullet, grunts, and snappers.

We loaded and returned to the ship to photograph specimens. After supper, duties were assigned to expedite collecting and diving procedures. Specimens caught by angling earlier that day were measured and dissected, and then night-lighting was begun at the anchorage (Fig. 5). Our first visitors were the kyphosid fish, *Sector ocyurus*, averaging about 0.3 m long, which appeared under our light by the hundreds. We subsequently learned that this Hawaiian species had been previously known from only one eastern Pacific

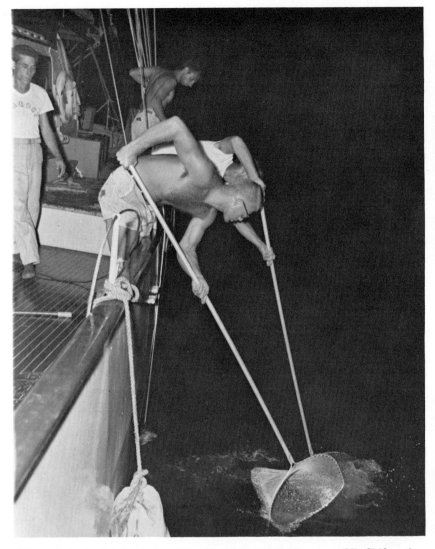

Figure 5. Night-lighting in Santelmo Bay, Islas Perlas, Panama (Vic Shifreen).

specimen. These fish were so thick that they rubbed against the light, and, as it was pulled slowly from the water, they followed it up so that the top layer of fish was lying on the backs of the fish below. Other species taken were mullet, anchovies, cornet-fish, crab zoea, mysids, and many other crustaceans, some in huge concentrations (*Argosy* Collection No. 5). Some of the

Figure 6. Surge channels in north cove of Bahía Piñas, Panama (Vic Shifreen).

party still-fished as well, and jacks (*Caranx crysos* and *C. caballus*), *Vomer*, and a species of *Lutjanus* closely resembling *L. griseus* of the western Atlantic were caught. Needlefish and a barracuda were seen about the boat but none was caught.

The heavy swells and uncertain quality of the water supply already started to take its toll, and by 2200 some of the scientific party turned in.

8 SEPTEMBER 1961

At 0800, we weighed anchor and departed Santelmo Bay toward Piñas Bay, Panama. *Sea Quest* refueled from *Argosy*'s diesel tanks and left for Cabo San Miguel with our three Panamanian guests whom we thanked for their help. From 1000 to 1100 we fished the HSPS, but the tow yielded absolutely nothing. Diving gear was readied and double-checked. In route, two rainbow runners (*Elagatis bipinnulatus*) of about 45 cm were taken trolling, and one was photographed. These are widely used by sport fishermen and native commercial fishermen who drift them for large marlin. A school of black skipjack was seen, but no other life was evident. A second tow was made with the HSPS, but this too produced little plankton.

The lovely lushness of Piñas Bay was visible by afternoon. Dense fog and rain clouds hovered over the tops of mountains that were about 200 to 250 m high. This dense rain forest (Fig. 6) is laved by heavy breakers that pound

Figure 7. *Argosy* at anchor, Bahía Piñas, Panama (Vic Shifreen).

the eroded scarps so characteristic of the Pearl Islands, but these seemed to me especially reminiscent of the northeast coast of Jamaica. At 1500, we passed between Morro Centinela and Morro de Piñas off the harbor of Piñas (see HO Charts 1410 and 1019) and anchored in about 10 m (Fig. 7).

Shortly after our arrival, we put the heavy, seemingly waterlogged lugger over the side and proceeded to poison a small rocky area near the village of Santa Dorotea on the north cove of the bay (*Argosy* Collection No. 6). A small, swiftly flowing stream washed into the bay, and casts of freshwater shrimp (*Macrobrachium*) could be seen scattered along it. Many fishes and one small octopus were taken from the rock pool. We learned later that here in June 1961 a shark had grabbed and disappeared with a small child in water less than 1 m deep.

Tony Provenzano and Dennis Paulson remained aboard *Argosy* to night-light at the anchorage. They subsequently reported that, as at Bahía Santelmo in the Pearl Islands, hundreds of the chub *Sectator ocyurus* crowded about the night-light. And lookdowns (*Vomer*) appeared in dense numbers to grab the anchovies attracted to the light. A number of the crewmen still-fished from the boats, and all specimens were eagerly preserved. Keeping the catch from the hungry crew was an ordeal in itself.

Figure 8. Left to right: Tony Provenzano, Clyde Roper, and Dennis Paulson preparing to set pyramidal trap and gill net in Bahía Piñas, Panama (Vic Shifreen).

During these activities, the rest of us had set out in the lugger and towed a 5-m balloon trynet back and forth across Piñas Bay. Only small fishes were taken (flounders, cardinalfishes, brotulids, scorpaenids, serranids) but none of these had been collected by us earlier on the trip, and so we were happy with what we got. Later we put over the pyramidal trap (Fig. 8) with bait and a light within, but found that the entrance holes were cut too high to permit most invertebrates to enter, and we got virtually nothing except crabs, which bit us badly.

9 SEPTEMBER 1961

The scientific crew packed up the lugger and headed south of Piñas Bay (Fig. 9) to the surf zone off the quaint fishing village of Jaqué (Fig. 10). Each

Figure 9. Mouth of Río Jaqué off Punta Jaqué, south of Bahía Piñas, Panama (Clyde Roper).

Figure 10. Village of Jaqué, Panama (Walter Starck).

Figure 11. Seining off Jaqué, Panama (Walter Starck).

time the waters yielded something different. Heavy surf along the beach some-
times prevented extensive collecting (Fig. 11), but the admixture of fresh
and salt water from the estuary (Fig. 12) yielded an interesting assortment
(*Argosy* Collection No. 9) of fresh and saltwater species. The local young-
sters assisted us by running after and collecting grapsoid crabs, digging for
worms, and generally joining in this bizarre merriment of beach collecting.
Later we went upstream into the rapidly flowing estuary, past the strong eddy
caused by the tidal and estuarine runoff. The sides of the river was very steep,
nearly 30°, and extremely difficult to collect in, but we could easily dive about

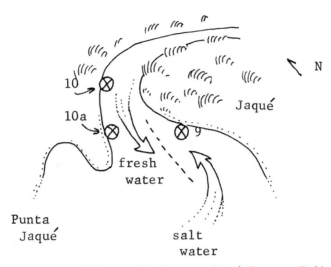

Figure 12. Location of *Argosy* collections near Jaqué, Panama (Robin Ingle).

and poison the rocky pools which contained many mullet and puffers (*Sphae-roides annulatus*). A good poison collection yielded blennies, gobies, snappers, and postlarval flounder and blennies. A tremendous number of oysters and barnacles studded the rocks along the water, and we cleverly managed to cut ourselves on all of them.

We returned to *Argosy* to photograph the specimens, work on gear, and hang nets. To our chagrin and amusement, we discovered that the net supplier in Miami had carefully rigged the float- and leadlines on the narrow ends of the 20-m seine (Fig. 13); fortunately, we had extra leads and floats, and we were able to rerig the net.

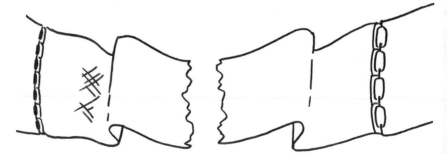

Figure 13. Twenty-meter seine as received from dealer (Robin Ingle).

Late that afternoon a shallow-water collection was made using spears and hand nets. In the evening, we night-lighted, set out the pyramidal trap, and the crew fished for specimens. Again, we had a good collection.

10 SEPTEMBER 1961

Today was our first attempt on this trip at collecting fishes using scuba gear, and we were quite apprehensive about the potential shark threat. We had been warned about the many shark packs in this area. One colleague in Miami had told us that he wouldn't dive in Piñas Bay even if his mother had fallen in. Each person was stalling to get his gear on so that he would not get into the water first. The day was overcast, the water quite turbid, the visibility poor, and the surge heavy. But we managed to get ourselves (Fig. 14) and the poison into the water, and we even obtained an excellent collection of fishes and invertebrates (*Argosy* Collection No. 13). Huge schools of a barred species of *Caranx* (*melampygus?*), tangs, and parrotfish were seen but not collected, as they seemed immune to or avoided the poison. After three hours of diving, we felt mixed relief and disappointment at not seeing our first shark. The rains had started, and we recharged our air tanks and wearily photographed specimens.

Night-lighting produced an assortment of fish postlarvae, crustaceans, and the usual predators snapping up the plankton and nekton attracted by the light. As the moon became brighter, we could detect what would prove to be a distinct but gradual decrease in abundance of organisms. Mullet of about 2 cm long frequently rested motionless under the light, curling the posterior part of the body at right angles which made them resemble leaves. A welcome addition to our collection, a 1-m barracuda, which looked like *Sphyraena ensis*, was caught on the bottom by Red Stuart, and several carangids were also taken and preserved.

11 SEPTEMBER 1961

The morning was shrouded in fog and drizzle, and because visibility was too poor to dive, we climbed the local mountains to collect land animals. We found many bats which dozed on the underside of banana leaves. The dense foliage was a welcome sight to us. At a height of about 100 m, Tony found an ovigerous hermit crab (*Coenobita*) chewing on the base of a banana tree. Katydids, moths, spiders, butterflies, and biting ants were everywhere, and monkeys and eleutherodactylid frogs could be heard in the distance.

Figure 14. Preparing to dive off Bahía Piñas, Panama (Vic Shifreen).

Dennis caught several birds with his BB gun, but the jungle's inhabitants generally hid from us. In the afternoon we poisoned a shallow riffle (*Argosy* Collection No. 15) in a tributary near Jaqué (Fig. 15). The creek could have been the Beaverkill River in the Catskills of New York, except that it yielded Panamanian fishes and not rainbow and brown trout. The catch in this shallow

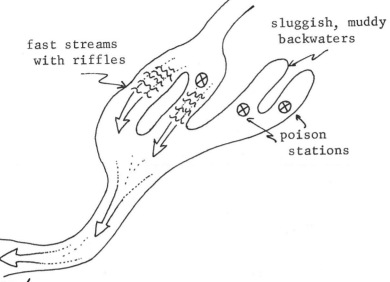

fast streams
with riffles

sluggish, muddy
backwaters

poison
stations

Jaqué

Figure 15. Location of *Argosy* Collection No. 15 in Río Jaqué, Panama (Robin Ingle).

riffle, with its quiet backwater pools, was a strange melange of fresh and saltwater species, consisting of gobies, several specimens resembling *Fundulus*, the killifish *Belonesox* which looks like a barracuda, and pipefish, cichlids, and atherinids which reminded us of the North American cyprinid *Rhinichthys*. After lunch, we set our multimeshed 100-m gill net across the mouth of Piñas Bay, and pulled it at 1730. Our entire catch consisted of two large houndfish, similar to *Strongylura raphidoma*, each about 1.5 m long. That evening we resumed our night-lighting, and the crew continued to catch specimens for our collections and for our larder.

12 SEPTEMBER 1961

Today was drizzly and cloudy, and we decided to brave the turbid waters and poison the Sentinal Rocks off the mouth of Piñas Bay. The larger of these two offshore islands, Morro de Piñas, closely resembled the heavily eroded rock we had seen at Isla de Rey in the Pearl Islands (Fig. 4). A smaller, flat-topped rock rose from the bottom at 20 m to about 4 m below the surface, and we quickly selected this site because of the blanket of fish covering the top of it. The water was starting to clear, and we made our best collection to

Figure 16. Moorish idol (*Zanclus cornutus*) speared by Walter Starck at Morro de Piñas, off Bahía Piñas, Panama (*Argosy* Collection No. 16), a first record for Panamanian waters (Walter Starck).

date, using poison, hand nets, and spears. Large amberjacks, snappers, jacks, and a huge manta, perhaps 6 m across, watched operations and scared us badly. Walter Starck speared a moorish idol, *Zanclus cornutus* (Fig. 16), one of three seen over the rocks. This species, common about Hawaii, was to our knowledge the first specimen taken in the eastern Pacific.

As the excitement increased over our good catch, we gradually realized that some of the shadows in the background, which we thought were jacks, were like a shark in outline, and that we were indeed surrounded by sharks. These were whitetip sharks, *Triaenodon obesus*, about 1.5 m long, and beautifully sleek and slender. They had been just out of our sight, paying no attention to divers until the poisoned fish started to react. Suddenly two sharks came in and immediately began groveling and picking up dead fish from the bottom. As I tried to extricate a dying triggerfish with my dip net from the rocks, one whitetip fearlessly rushed in and grabbed the triggerfish from within the net and made off with the fish *and* the net. But they didn't bother us further, and we were too excited about the collection to worry about the sharks.

This was the day we were to be thankful for our survival training in the divers' school in Miami. At about 20 m, Tony's regulator began to malfunction, and with each breath he inhaled water. His diving buddy, Clyde Roper, fortunately was by his side, and upon Tony's signal that he had no air, Clyde and he buddy-breathed to the surface on Clyde's regulator. It was a close scare none of us has forgotten. Later, we found that a tiny sand grain had been caught in the regulator valve, preventing its complete closure.

After lunch aboard *Argosy*, we pulled our gill net and found, over its 100 m length, one grunt and one chub (*Kyphosus*). But we also found several dozen holes averaging about 0.5 m across. The gill net had been torn up badly, probably by sharks eating the gilled fish.

The weather worsened, and at about 1330 we decided to postpone that afternoon's planned diving trip. Instead, we took *Argosy* offshore to dredge for bottom invertebrates with our new, but as yet untried, homemade winch (Fig. 17). After a trial haul at 40 m, in which we obtained a single olive shell, a handful of worm tubes, and a tree branch, we tried again, this time at 80 m. Suddenly, *Argosy*'s fathometer broke down, and we did not know that the bottom was shoaling rapidly. We then realized that we had unexpectedly anchored *Argosy* with our dredge. After several unsuccessful attempts to back down, the winch clutch-plates failed to hold, and we lost the dredge.

Disappointed, frustrated, and weary from an already complete day of hard

Figure 17. Trawl winch being operated from stern of *Argosy* by *left*, Roper and *right*, de Sylva (F. G. Walton Smith).

Figure 18. Mako shark (*Isurus oxyrinchus*) caught by Mr. Alfred C. Glassell, Jr., off Bahía Piñas, Panama. *Left to right*: Mr. Glassell, Paulson, Provenzano, Roper, de Sylva, Starck (Vic Shifreen).

knocks, we returned to Piñas Bay and awaited the return of Mr. Glassell from his day of fishing. He had returned with a boatload of fish of every description. He had taken another of his many beautiful sailfish and some bonitos, tunas, jacks, and a 148-kg female mako shark, *Isurus oxyrinchus* (Fig. 18). Also on

Figure 19. *Sea Quest* returns with a dusky shark (*Carcharhinus obscurus*) to Bahía Piñas, Panama (Vic Shifreen).

deck was a dusky shark, *Carcharhinus obscurus* (Fig. 19). We measured and dissected them, and were happy to get our hands on this selection. That night we set several shark hooks from the stern of *Argosy*, and we all went to bed, in hopes of what *had* to be a better day tomorrow.

13 SEPTEMBER 1961

Today we departed Piñas Bay in search of a guyot, or seamount, which was reportedly about 130 km west southwest of Piñas at about 07°N, 79° 05′W (HO Charts 1176, 1019). The existence of this guyot had not been confirmed recently by other vessels, but we were going to try to find it from the chart, without a fathometer and using only the stars, and to try to anchor on a mountain top of 22 by 32 km charted at from 80 to 100 m deep surrounded by depths of over 1800 m. On our way we towed the high-speed plankton sampler for an hour and obtained a good collection of siphonophores, crab megalops, stomatopod larvae, fish larvae, and shelled mollusks. A second tow also yielded a fine collection, including a crustacean which none could assign to any major group and which excited everyone, even the iguana we had picked up in Panama. We immediately put over the sampler again, this time abeam of *Argosy*. She rode beautifully, submerged about 1 m. As we watched, in horror, a huge log loomed on our starboard side. The sampler hit head on, the thin towing cable parted like thread, and that was our last experience with the high-speed plankton sampler.

With the determination of Tennyson's Light Brigade, we sailed onward toward our seamount. The weather was excellent and the seas perfect. The long haul gave us time to reflect on our successes, but particularly on our misfortunes. In six days:

(1) *Argosy's* radar had broken down.

(2) *Argosy's* air conditioning ceased functioning, but eventually was repaired.

(3) The AC-DC converter we had purchased in Panama to run our air compressor from *Argosy* failed to operate, although later we were able to install a DC motor directly onto the air compressor.

(4) A large electric reel for deepwater fishing, built at a cost of nearly $3,000, never worked and could not be repaired by us.

Figure 20. Towing dead sharks out to sea off Bahía Piñas, Panama (Vic Shifreen).

(5) A shipment of light bulbs and plastic lights that were to be used for nightlighting and in baited traps had not arrived in Panama in time for our departure.

(6) Two of our diving regulators had malfunctioned after being used twice.

(7) Our bottom dredge was lost in the rocks.

(8) The gill net was torn up badly by large fish, and most of it had to be discarded.

(9) Our seine had been rigged wrong by the supplier.

(10) The deep-sounding recording fathometer had broken down.

(11) Our winch was not operating properly.

(12) The rheostats for our nightlights were not operating.

(13) Our plankton sampler was lost.

Only those who "go down to the sea in ships" can appreciate the frustration but necessary acceptance of such problems. No amount of careful planning can overcome tough luck, and no amount of pretesting or prechecking of equipment seems to prevent subsequent breakdown. John Steinbeck, in his "Log from the *Sea of Cortez*," showed a special understanding of the demon

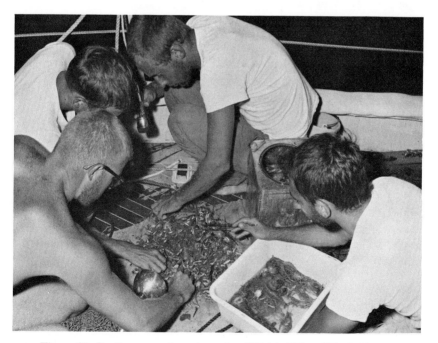

Figure 21. Sorting a trawl catch made off Bahía Piñas (Vic Shifreen).

Figure 22. Captain Arved J. Rosing, master of *Argosy* (Walter Starck).

god of the sea who carefully arranges hard luck for mariners. What we *did* have aboard was a year's supply of humor, but we nearly used this up the first week at sea.

By 1545, Captain Rosing's navigation had put us in the general vicinity of the guyot (Fig. 22). Since the fathometer was not working, the crew took

Figure 23. A young yellowfin tuna comes aboard *Argosy* off Panama (Vic Shifreen).

repeated soundings, using four 2-kg sash weights on the winch, which by now was again operative. As we continued sounding, we observed an increasing abundance of sea life, including six Ridley turtles which were captured and measured by Dennis Paulson, porpoises, flyingfish, brown and redfooted boobies, and shearwaters. We put out a trolling line and, as fast as we could get a feather in the water, we hooked several skipjack tuna (*Katsuwonus pelamis*), but lost all because of the vessel's speed.

We had been traveling over depths, according to our charts, which exceeded 1800 meters, but we now knew that we were very close to the guyot; the water was alive with plankton, fish, birds, everything. Compared to the waters to our east, these upwelled waters, rich in nutrients, were supporting a luxuriant food chain. Suddenly our flashing fathometer, which recorded above 180 m, began to flash repeatedly, and we knew that Captain Rosing had found his submarine mountain. In the distance we could see Mr. Glassell battling a fish. When he returned to *Argosy* later, Mr. Glassell had a 3-m sailfish, two yellowfin tuna (*Thunnus albacares*) (Fig. 23), six or seven black skipjack (*Euthynnus lineatus*), some beautiful dolphin, and two sharks which we were unable to identify.

Figure 24. Silky shark (*Carcharhinus falciformis*) caught at seamount 130 km off Panama. Packs of these roving sharks swarmed about *Argosy* as she hove to (Walter Starck).

About 1700 we managed to anchor on the guyot, and by 2000 hours we were able to night-light and bait up our shark hooks. Using cut skipjack, Walter Starck caught ten of what we thought were silky sharks, *Carcharhinus falciformis* (Fig. 24), from 1.5 to 2 m, as fast as he could get the line in the water. As he hooked them, I shot them in the head with our Model 1903 Springfield .30-06, which would prove so useful for sharks throughout the trip. The sharks were measured, dissected, and the jaws and stomach contents preserved. In the meantime, Red Stuart, fishing from *Sea Quest*, caught eight more. Although we were 75 km from land, sharks were still everywhere. They had been attracted either to the small fishes near our night-light or to the light itself; we believe it was the latter because they moved in within seconds after the 100-watt light was dropped overboard. It was surprising to us to see that these sharks wasted no time in circling but made a beeline for the light when it was put in the water, becoming apparently frenzied as they approached the light. Similarly, when chunks of black skipjack were put on a shark hook, there was no preliminary pass but rather a direct attack on the bait. Their fearlessness was frightening. We were more than glad that we had been dealing with the "more gentle" whitetips (*Triaenodon*) when we were diving off Piñas Bay. Gradually, as more sharks were shot and landed, the readiness of the others to take a bait diminished, and within two hours the sharks had disappeared.

The night was black and the sea dead calm. During this period the less excitable members of the crew and scientific party avidly night-lighted from the other side of *Argosy*. They were eventually joined by everyone, and, after five hours of night-lighting, we had produced a phylogenetic wonderland of

Figure 25. Young dolphin (*Coryphaena*) dipnetted at the seamount off Panama (*Argosy* Collection No. 17) (Walter Starck).

goodies, including squids, shrimps, crabs, tunicates, an assortment of fishes, and a Galápagos petrel which, blinded by the night-light, flew into the dip net. Five gempylids, which appeared to be *Nesiarchus*, many luminous lanternfish, needlefish, several young tunas, flyingfish, postlarval squirrelfish, and the rare stromateoid fish *Cubiceps* were taken (*Argosy* Collection No. 17). Several dolphin (*Coryphaena hippurus*) of about 20 cm were taken, two of which were kept alive in our circulating shipboard aquarium (Fig. 25).

14 SEPTEMBER 1961

Heartened by last night's excitement, we had trouble sleeping and awoke at 0530. But we took one look at the drizzling overcast day and went back to bed. After breakfast, *Sea Quest* trolled in a radius of several km around *Argosy*. Later that morning the rain stopped, and we photographed specimens taken the previous night. Shortly thereafter we put over a modified 5-m balloon trawl which we attempted to fish as a mid-water trawl with 180 m of cable out. We trolled feather jigs and dipnetted seaweed in route. The trawl yielded nothing but a few tunicates after an hour's tow, and the only fish taken by *Argosy* was a handsome 15-kg dolphin hooked by Tony Provenzano (Fig. 26). Later that afternoon we filled our air tanks and readied our diving gear for the next shore station. In the moonless evening we reanchored on the guyot and night-lighted. Walter Starck and Dennis Paulson put over a light and shark hooks and landed eight more sharks within a half hour. Night-lighting was excellent; our first catch, in three passes of the dip net, was eleven small Pacific sailfish (Fig. 27) which virtually swam into our net. They were

Figure 26. Tony, with friend caught at seamount off Panama (Walter Starck).

kept alive for several hours in our shipboard tank, but repeatedly rammed their bills into the corners of the starboard aquarium, and eventually all died. Activity was much less than on the previous night, and fewer fishes were seen but, to the delight of Clyde Roper, many squids were taken, and some were maintained in the port aquarium.

About this time we ran into our first group of yellow-banded sea snakes, *Pelamys platurus*. These deadly creatures swim all over the eastern tropical Pacific close to land, yet we had not encountered them up to now except in

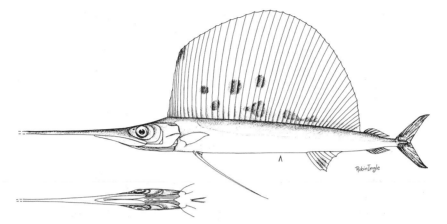

Figure 27. Young sailfish (254 mm fork length) taken by dip net and night-light at seamount off Panama (*Argosy* Collection No. 18) (Robin Ingle).

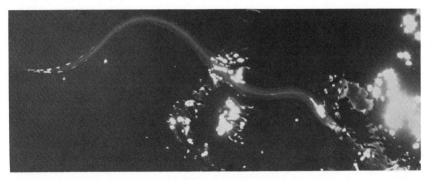

Figure 28. Yellow-banded sea snake (*Pelamys platurus*) attracted to night-light of *Argosy* over seamount off Panama (Walter Starck).

stories, and we were surprised at their appearing so distant from land. They were readily attracted to the light (Fig. 28), appeared momentarily, disappeared beneath the surface, and again materialized as if from nowhere. Their swimming agility, either forwards or backwards, was incredible. Several were easily dipnetted, but at one point, while I was dipnetting from a partially submerged platform abeam of the ship, the ship rolled as we drifted through a school of several dozen. I was up to my chest in sea snakes to the amusement of my companions, but fortunately the danger passed quickly as they drifted by.

Clyde and I rigged the pyramidal trap, baited it with several chunks of skipjack, and dropped it to a level of 50 m on the winch cable. One hour later we retrieved the remains of the trap. The sharks had ravaged the trap with these small chunks of skipjack—there remained only splinters of heavily slashed wood and the galvanized eye bolt to which the trap had been attached.

At 2215 under a drizzly black sky, we weighed anchor and departed for Punta Utría, Colombia.

15 SEPTEMBER 1961

It was still raining when we saw the fog-shrouded hills of the Colombian coast. We could see Punta Utría in the distance. To the south was Bahía Solano, and we arrived within the hour. This lovely, quiet arm of the sea clothed in mist and dense rain forest is where the Pan American Highway was eventually to link up the Americas and is also the terminus of one route of the proposed sea-level canal. We saw no one here, and *Argosy* coasted down the bay to our anchorage near the tiny village of Cuidad Mutis, a meeting place for about eight hundred persons including workers on the Pan American Highway, a few fishermen, and some farmers. Bahía Solano is a lovely, placid spot and, we learned later, is fed by a number of streams which carry gold dust from the neighboring mountains. We also learned that the leading product of the rich Chocó province is platinum.

As the weather cleared and the rains desisted, we dropped anchor, and, to our surprise, found that at only several hundred meters from the bay's shore we were anchored in 100 m. We launched the lugger and explored the reddish brown sand beach. Above us were numerous waterfalls cascading over basaltic boulders with many tiny rivulets which permeated the narrow beaches.

Some fishermen in a dugout canoe had just hooked about fifteen chubs (*Kyphosus*) of about 1 kg each and a red grouper (locally *pargo colorado*) which they had taken at the south end of the bay. They claimed the water was largely too deep for effective fishing, but that fishes were abundant. They noted, casually, that sharks were common.

Our group poisoned a small tide pool (Fig. 29) and estuary (*Argosy* Collections Nos. 19–21) which produced eels (*Myrichthys*), pomacentrids (*Abudefduf saxatilis* and *A. analogus*), mugilids, blennies, gobies, holocentrids, an unusual species of the serranoid fish *Rypticus*, and several palaemonid shrimps. As the poison drifted along the beach, a school of small threadfin (*Polydactylus*) was hit, and several pompano (*Trachinotus rhodopus*) were seen but could not be collected.

Later that evening we put ashore in the lugger to Ciudad Mutis and sampled the fine Colombian beer and, what was to prove deadlier than the sharks, Colombian *aguardiente*—fire water.

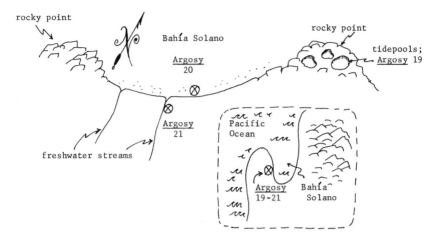

Figure 29. Location of *Argosy* collections at Bahía Solano, Colombia (Robin Ingle).

16 SEPTEMBER 1961

The morning was overcast but bright, and the rains of the past day had riled the bay water. We traveled north with our diving gear to Cabo San Francisco, the promontory guarding the entrance to Bahía Solano. At a rocky point just outside the bay, we found an area that looked ideal to poison. We were hesitant to enter the water, however, because yesterday we had seen a species of whitetip shark which we thought possibly might be *Carcharhinus albimarginatus*, and today we saw another species of shark. We were loaded with spears, knives, tanks, nets, shark poles, crowbars, jars, poison, and cameras, and each of us was using this cumbersome gear to full advantage in hopes that someone else would get in the murky water first. Tony Provenzano, burdened with about 20 kg of gear, finally jumped into the water. But within what seemed to be milliseconds he was back up and, despite the weight of all his gear, into the boat. As he hurdled the gunwhale in a lightning shot, he blurted that he had landed on the back of a *Carcharhinus albimarginatus* —whitetip. Needless to say we dived elsewhere.

Eventually we found a spot amidst large boulders that had been pounded and smashed by heavy weather but still provided a home for marine life. A good collection was made of fishes and invertebrates, including a small octopus with beautifully banded arms and body. Many larger fishes were seen but could not be collected with poison. They simply did not react to it. Among these were two large specimens of *Caranx hippos*, many *Lutjanus argenti-*

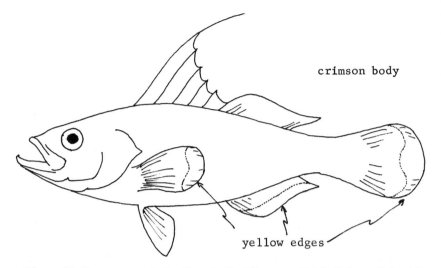

crimson body

yellow edges

Figure 30. Snapper caught by Captain Red Stuart at Bahía Solano, Colombia (Robin Ingle).

ventris, plus several specimens of *Haemulon scudderi* or *maculicauda* and *Anisotremus dovii*, with a yellow rear, several specimens of *Kyphosus* and *Sector* (the latter the rare kyphosid fish first seen at Islas Perlas), numerous blue-ringed *Cirrhitus rivulatus* perched about the rocks, a few *Chaetodon humeralis*, *Holocanthus passer*, both young and adult, a *Xesurus* in a yellowish orange phase, and an unidentifiable species of *Microspathodon*. Also seen but not collected were two species of *Abudefduf* (*saxatilis* and *taurus?*), also both Atlantic forms, *Bodianus diplotaenius*, both male and female, *Chromis*, *Fistularia*, *Sphaeroides* sp., and a beautiful species of *Acanthurus* with a white bar on the caudal peduncle. Two medium-sized spiny lobsters (*Panulirus gracilis*) were seen but not collected. Toward the end, Tony picked up a large, live stargazer (*Astroscopus*) with a pair of forceps. In all, it was our best and most representative station to date (*Argosy* Collection No. 23).

After lunch we anchored the lugger just offshore, with Walter staying aboard, and the rest of us waded in to explore a beautiful waterfall. The falls tumbled over clean black bedrock, free from the usual slippery algae, and we could climb with ease. The view was magnificent. We found marine snails similar to our West Indian *Neritina* about 30 m up the falls, but this was as far as we went, for at that moment we heard Walter yell. We looked in horror to see our boat wash into shore and about to flood. We scrambled down the falls only to see the boat broach to the long rollers. It had broken its fore and aft anchor lines. The grounded boat flooded but did not capsize. Equipment sloshed back and forth. We could see our diving gear and some specimens floating out of the boat, but we could not retrieve anything because we had to hold the heavy boat to keep it from shoaling further. Finally, with the aid of two of the many ubiquitous natives, we got the boat off the shoal, refloated it, bailed out, and eventually most of our equipment was taken ashore. A mask, a snorkel, two new species of hermit crabs (subsequently taken elsewhere on the cruise), a banded octopus, a rare starfish, some fine coral, and a sea urchin, plus our pride, were lost. Despite our attempts to start the flooded engine, it simply would not budge. At 1300 we started rowing and by 1620 we arrived soggily at *Argosy*.

We slept until 2100 while the chief engineer successfully dried the lugger's engine. We departed offshore of Cabo San Francisco for night work aboard *Argosy*. Our 5-m balloon trawl was fished in mid-water about 15 km offshore as we towed our 1-m plankton net at the surface. The plankton tow was not particularly rich, but we were pleased that our homemade "mid-water trawl" had worked at all. Hatchetfish, cutlassfish, salps, tunicates, penaeid and caridean shrimp, eel larvae, and ctenophores were but part of our catch. Night-lighting at midnight yielded nothing, and we all turned in.

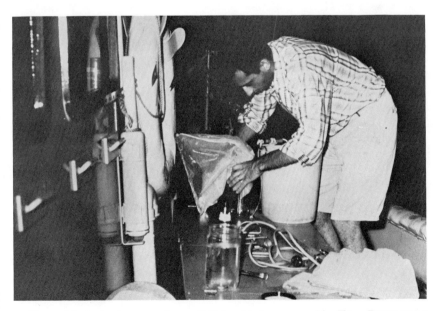

Figure 31. A plankton tow from off Colombia is preserved by Tony Provenzano (Walter Starck).

17 SEPTEMBER 1961

As we pushed south all morning toward Buenaventura, Colombia, past Nuquí and Cabo Corrientes, we rerigged gear, charged tanks, and photographed specimens. Despite the daylight, Red Stuart dipnetted three small cutlassfish (*Trichiurus*) from the surface. They were merely lying there and made no attempt to avoid the boat, and all were in excellent condition. A small Ridley turtle was also captured. After supper two drags with a 5-m

Figure 32. Preparing specimens for shipment to Miami from Buenaventura, Colombia (Vic Shifreen).

Figure 33. Marketplace at Buenaventura, Colombia (Vic Shifreen).

trynet (*Argosy* Collection No. 26) yielded a fine assortment of estuarine, mud-bottom animals, including flatfishes (*Symphurus* and *Paralichthys*), opisthognathids, scorpaenids, several species of serranids and batrachoidids, and a variety of penaeid shrimps. During our bottom trawl we towed our 1-m plankton net which yielded excellent collections of larval fishes and copepods (Fig. 31). A second tow was quickly set, but the *Argosy*'s steward, unaware of the men of science in action, threw the entire contents of the garbage can into the net. After unsuccessfully attempting to extract plankton from the chicken bones, coffee grounds, and egg shells, we discarded it and set the plankton net for the third time. The last haul was without garbage, and we obtained a nice series of tripletail (*Lobotes*) and dolphin (*Coryphaena* sp.), as well as mackerel (*Scomberomorus*), flyingfish, and jacks.

18 SEPTEMBER 1961

After spending the night in the harbor of Buenaventura, we moved upstream into the estuary of the Río Dagua. The muddy, brown outflow was distinct from the clearer waters we had traversed to the north. The lowlands of the Chocó were laced with extensive mangroves, and there were a few coconut palms on the higher ground. At about 0930 we arrived at Buena-

Figure 34. "Gonzalez," the three-week-old Colombian ocelot, mascot of *Sea Quest* (Vic Shifreen).

Figure 35. A moment's rest is enjoyed by all at El Campín, Buenaventura, Colombia. *From left*: Roper, Starck, Bill Saenz, Provenzano, Eric Saenz, de Sylva, Paulson, Luis Borda (Vic Shifreen).

ventura, said to have the heaviest annual rainfall in the world (nearly 800 cm), and it felt that way. After clearing customs, we met William Saenz, a former University of Miami employee and now assisting the Colombian fisheries program, and his colleague Dr. Luis Ortíz Borda from the Universidad de Bogotá. After picking up fuel, food supplies, electrical parts, and many other items, we shipped our collections home (Fig. 32) via Grace y Cía., thanks to the capable assistance of Señor Gerardo Marquez and Captain Robert Early. That evening we were invited to participate in the annual Beethovenfest and cultural exchange occurring at the Universidad El Campín in Buenaventura (Fig. 35).

19 SEPTEMBER 1961

After less than two weeks at sea, our supplies were getting low, thanks largely to the ravenous appetites of the scientists. We therefore spent the day in Buenaventura trying to replenish our supply of food and water. Later in the day we all visited the renowned shrimp fleet based at Buenaventura (Fig. 36) and examined the extensive facilities of Industrias Pesqueras, S.A., which is one of six local shrimp industries. The shrimp were huge—13 to 26 per kg,

Figure 36. Shrimp fleet at rest in Bahía de Buenaventura, Colombia (Walter Starck).

and there were reports of even larger shrimp. This operation employed perhaps one hundred people in an immaculate, efficient setting. Shrimp handlers,
peelers, and packers seemed to be everywhere. A large quick-freezer was in
operation for glazing prior to shipment to the U.S. In 1959, 80 boats fished
from this port, but we were told that 150 shrimp boats could easily fish from
here. Some fishermen had even made enough money in six months of shrimp
fishing to pay for their boats.

20 SEPTEMBER 1961

We finally obtained our cooking gas and departed Buenaventura at about
noon. Today was the first day we had sufficient wind to sail, and we all helped
enthusiastically with the canvas (Fig. 37) as *Argosy* headed for Gorgona
Island. Few birds were seen, although some porpoises followed us out of the
harbor. We sailed on, with *Sea Quest* going ahead to fish in route. We were
anxious to reach Gorgona Island, the island reputed to have one poisonous
snake per square meter. We checked our gear, filled air tanks, and spent the
rest of the day trying to read James Bond novels in the gloom of the haze and
rain.

Figure 37. Hoisting sails off Bahía de Buenaventura (Vic Shifreen).

21 SEPTEMBER 1961

The dense stands of coconut palms along the beach on Gorgona were shaded by the high, volcanolike cliffs hanging over them. Gorgona and her sister rocklet to the south, Gorgonilla, presented a peaceful sight to us as we put over the lugger at 0700. A two-hour intensive poison station would yield fine results. This station was unusual to us because the steep underwater slopes of the island permitted pelagic fishes, such as whitetip sharks, rainbow runners (*Elagatis*), sierra mackerel, and black skipjack, to come in close to us while we were poisoning coral-reef fishes. The contrast was something to behold, until we realized that our pelagic curiosities were eating a considerable number of the fishes we had poisoned, and we unsuccessfully tried to scare them away. We were also out of luck in getting rid of the 1.5- to 2-m whitetip sharks. (*Triaenodon obesus*), which also frequently took the dead and dying fishes from under our noses; but, nevertheless, we managed to make a fine collection (*Argosy* Collection No. 27). Several species of snappers were seen and photographed (Figs. 38, 39) but were not affected by the poison, nor were triggerfish, grunts (*Haemulon maculicauda*), large red-and-green parrotfishes, jacks with bright blue tails, the wrasse *Bodianus diplotaenius*, and large schools of goatfish (*Mulloidichthys*). Thousands of holocentrids were seen lurking among the rocky crevices, and a large series was collected with

Figure 38. Grunts and goatfish at Gorgona Island, Colombia (Walter Starck).

Figure 39. Snappers and grouper at Gorgona Island (Walter Starck).

Figure 40. A rare cirrhitid (*Oxycirrhites typus*), a new record for Colombian waters. Speared by Walter Starck at Gorgona Island (*Argosy* Collection No. 27) (Walter Starck).

our nets. The cirrhitid fish *Cirrhitus rivulatus* was common, and we finally captured several of many seen of the rare cirrhitid *Oxycirrhites typus* (Fig. 40) which were lurking about gorgonids. Several more moorish idols (*Zanclus*) were schooling with butterflyfish. Scattered about the talus slope of Gorgona were the burrows of the sand eel, *Malacanthus*, but none was taken.

Toward the end of the collection, Tony's regulator again failed to function, and he and Clyde buddy-breathed from a depth of 15 m. This time the prompt lifesaving technique almost seemed routine, but we again thanked ourselves for the caution we might have once considered excessive.

After we had lunch, *Argosy* returned to the shore of Gorgona to allow us to snorkel and poison among the large boulders of the talus slope surrounding Gorgona. The many crevices served as an ideal habitat for holocentrids, serranids, lutjanids, and, to our dismay, morays of about 1 m which became

Figure 41. Mr. Glassell and his 180-lb. (82-kg) sailfish taken west of Gorgona Island (Vic Shifreen).

exceedingly bold and nasty from the combined effects of the poison and the sudden availability of numerous dead and dying fish. Many angelfishes (*Holacanthus passer*) were seen and three more *Zanclus* came by, one of which was speared with a multipronged spear by Walter Starck. A few large jacks, *Caranx hippos*, blue-spotted balistids, and orange-bellied balistids were seen. A large manta ray came in close to our operations, followed by several amberjacks (*Seriola*), which swam remora-style closely beneath the manta's wings.

Later we saw a huge school of the Indo-Pacific species *Kuhlia sandvicensis* (?), characterized by the black-banded, deeply forked tail. To our knowledge, this species had not been seen previously in the eastern Pacific. Of interest also were two large milkfish, *Chanos chanos*, also believed only to inhabit areas about Hawaii and the western Pacific.*

Dennis and Luis Borda had already gone ashore to collect and poison tide pools, and we were concerned for them because of the rumors of poisonous snakes. Eventually, despite the warnings of a few natives in dugout canoes that anyone going into the bush would never return, Dennis and Luis did emerge with a bag of lizards, birds, and a couple of bushmasters for the collection.

Late that afternoon Mr. Glassell returned on *Sea Quest* with a nice 180-lb.

---

*One specimen of about 20 kg was subsequently caught (about 1966) off western Mexico.

(82-kg) sailfish (Fig. 41) and two yellowfin tuna. The sailfish gonads, as those of the other sails taken earlier on the trip, showed no signs of spawning.

Gorgona Island is a penal colony of the Colombian government, and the dense jungle with its poisonous snakes and surrounding shark-infested waters make it unlikely that anyone would try to escape from it. The prison guards confirmed this and added that the island harbored an extinct volcano that was the home of a huge creature which the inmates greatly feared. On several occasions it was reported that a chunk of meat had been tethered to a stake with a chain, and when the prison guards had returned to the spot, the chain had been broken. So no one really knew *what* was there, but on this island time was lost, and imaginations were rich. A delegation from the prison came out to our ship, and we showed our visitors our facilities and collections. We joined them in some *aguardiente*, and they invited us ashore for dinner. Tony, Clyde, Bill and Eric Saenz, and Luis Borda went ashore, and the rest of us stood by the ship and night-lighted, while the crew still-fished for four hours. In spite of a heavy, intermittent rainfall, we obtained what was one of our best collections. Between downpours we could see a tremendous amount of plankton drifting by, and any time a fish was dipped up or caught on hook and line, the water came alive with phosphorescence. And at one point something that looked like a 2-m torpedo, glistening with phosphorescence, zoomed into our night-light and disappeared. It may have been a porpoise, but its high speed and ominous appearance shook us up considerably.

Figure 42. Armed with multipronged spears, de Sylva and Starck prepare to dive along the slope of Gorgona Island (Vic Shifreen).

22 SEPTEMBER 1961

Despite another rainy, overcast day, Walter and I went ashore and dived along the steep slope east of Gorgona (Fig. 42). At a depth of 20 m we found the peculiar sand pyramids of garden eels (Heterocongridae), with the inhabitants swaying in the current and picking bits of plankton from the surrounding waters. As we approached them, they slowly retreated into their burrows, popping back out after we had passed over them. We managed to take two (*Argosy* Collection No. 33) with the multipronged spear, but because it lacked barbs, we lost several more. These eels held tenaciously to

Figure 43. Mr. Glassell fights a nice striped marlin on ultra-light tackle off Gorgona Island (Vic Shifreen).

their burrows, and were virtually impossible to extract. Even by squirting rotenone and a narcotic into their burrows, we were unsuccessful in capturing any more.

Hundreds of labrids and a small species of *Serranus* were seen, and a school of about thirty specimens of *Elagatis bipinnulatus*, the beautiful yellow and blue rainbow runner, swam around us while we were on bottom. Wide areas of gray sand bottom were interspersed with a few boulders or patches of

Figure 44. Red Hagen bills a sailfish off Gorgona Island (Vic Shifreen).

Figure 45. A 325-lb (148-kg) black marlin taken by Mr. Glassell off Gorgona Island is examined for stomach contents (Vic Shifreen).

sponge which the few butterflyfish, angelfish, and labrids desperately seemed to seek as shelter. At about 30 m, I speared a large *Caranx* (*melamypgus?*), but it swam off with my spear, leaving a trail of blackish green blood behind it. I had forgotten how color changes with depth.

As we returned to the ship, we met the group which had returned from their night of revelry ashore. They had eaten a marvelous meal, been received warmly, and had spent the night sleeping in jail. They had all picked up the beginning of colds a few days earlier, however, and were feeling miserable. Several of us made some additional dives to poison or to spear specimens, but visibility was extremely poor. We also put out a six-hook trotline in 30 m, but hours later we had nothing but cleaned hooks.

We had planned to leave for Tumaco, on the mainland, at 1500, but Mr. Glassell excitedly returned with the report that he had hooked and lost a beautiful black marlin estimated at over 250 kg. We had all heard of large marlin in this area, and there was an unconfirmed report that we had been given while we were in Buenaventura of a 109-kg sailfish taken northwest of Gorgona. If confirmed, this would have been a world's record. So we decided to stay another day and hope that the big marlin would return via the *Sea Quest*. That afternoon Mr. Glassell had taken striped marlin (Fig. 43), sailfish (Fig. 44), and a 325-lb. (148-kg) black marlin (Fig. 45), which to us was no small fish. This female also showed no signs of spawning. Caught also was a whitetip shark, *Carcharhinus longimanus*, estimated at 115 kg, which we had not yet encountered during our dives, but which we had seen from afar. This nasty species, fortunately, is found over deep waters.

By this time, we were all feeling pretty sick—Clyde had a cold and had also dropped a knife into his foot, Walter had strep throat, Dennis and I had a virus, and Tony was coming down with a virus. From the way our stomachs felt, we had probably taken on bad water at Buenaventura, and most of the expedition members were down by nightfall.

23 SEPTEMBER 1961

In spite of a long night's sleep, we still felt ill today. The rain and permeating dampness did not help our well-being, and both *Argosy* and *Sea Quest* had begun to look like hospital ships. We sluggishly checked our diving gear, filled our air tanks, and moved *Argosy* toward Gorgonilla, Gorgona's smaller sister island at the south end. Despite torrential rains, Mr. Glassell valiantly spent the day trolling offshore of Gorgona.

After lunch, Walter and I donned our wet suits and spear guns and pointed the lugger for Gorgonilla. Despite the weather, the incredible beauty of this island was striking, with its sandy beach bordered by densely packed stands of coconut palms rising sharply up volcanic cliffs. Thick mats of strangler vines interwoven with large balsa trees inland were reminiscent of Piñas Bay and Islas Perlas. We anchored the lugger off the steep beach and swam ashore to poison the rough surf zone, but we obtained absolutely nothing here. Numerous reddish ghost crabs skittered along the beach and, when chased, ran lemminglike into the heavy surf. Several of these plus a large hermit crab (*Coenobita*) were collected for Tony. We returned to the lugger, donned our gear, and dove to 29 m along the 35° slope of the island. Nearer the surface were boulders and cobbles, but sand and silt predominated below. Two spiny

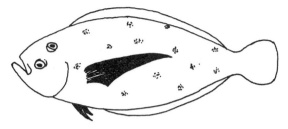

Figure 46. Flounder seen at 29 m at Gorgona Island (Robin Ingle).

lobsters (*Panulirus gracilis*) of 0.5 and 1 kg were speared, and a flounder of about 0.5 m with a black pectoral fin (Fig. 46) was chased but lost. Shallower inshore, at about 3 to 9 m deep, a coral garden, perhaps covering an acre, was certainly among the most gorgeous we had seen. Although it was composed of only a few species (largely several species of *Porites* and *Montepora*), it sheltered a superb variety of fishes and invertebrates whose colors, in spite of the poor visibility, were remarkably vivid. Porcellanid and grapsoid crabs, many types of snapping shrimp, holothurians, and seemingly thousands of ophiurians per square meter were jammed into this impregnable niche. Browsers and invertebrate-eaters, such as triggerfishes, parrotfishes, pomacentrids, squirrelfishes, surgeonfishes, porcupinefishes, and wrasses, combed the area continuously.

By 1630 hours, an eerie, incandescent greenish glow seemed to pervade the coral and surrounding waters, and because the light had become very dim, we surfaced and returned to *Argosy*. Mr. Glassell, aboard *Sea Quest*, had seen several marlin but the weather was rough and fishing was poor. A female whitetip shark (*Carcharhinus longimanus*) of about 115 kg which Mr. Glassell had caught carried two near-term pups about 0.3 m long. The pups had no visible teeth, but the light-tipped fins were already evident.

Argosy moved back into her sheltered cove and anchored. We tried night-lighting for a few hours, and got some flyingfish, mullet, goatfish, and a small *Physalia*, but *Argosy* was rolling badly in the heavy swells, so we moved south of Gorgonilla and headed toward the mainland metropolis of Tumaco. During the late evening we made a fine bottom collection with the 5-m trynet at a depth of about 90 m of water (*Argosy* Collection No. 38) which yielded an assortment of estuarine and deep-slope fishes, indeed a strange combination to us. Simultaneously we towed our 1-m plankton net. We did not get any garbage this time, but as we retrieved what was a really rich haul, I started to put my hand into the net to shake out some weed we had amassed, as is usual with a plankton tow. Suddenly Clyde turned on his flashlight and yelled. A sea snake (*Pelamys platurus*), about 1 m long, was packed in the weed, and Clyde had spotted it just in time.

Figure 47. From the harbor of Tumaco, Colombia, several scientists prepare to depart for home (Vic Shifreen).

24 SEPTEMBER 1961

As we approached the southern part of the Chocó—the Pacific lowlands of Colombia—we could see how different this was from previous areas. Stretches of impenetrable mangrove guarded the entrance to the little fishing village of Tumaco, with extensive, shallow tidal flats as far as one could see. Small sail-powered dugout canoes rode the interweaving estuaries feeding Ensenada Tumaco, and cast nets and gill nets could be seen in operation. After clearing with the harbor authorities, we went ashore in *Sea Quest* to discharge part of our group—Dennis, Tony, Walter, Bill and Eric Saenz, and Luis Borda (Fig. 47). Tumaco's residents are a proud people, but their isolation, the rainy climate, and the lack of modern facilities pose great inconveniences for them. We headed immediately for the fish market to purchase specimens, but all we could find were some unrecognizable split and dried carcasses which may have been fish. Apparently there is considerable fishing activity by the natives, and a well-constructed beach seine was being repaired near the market. We had seen some mullet and pompano being taken in gill nets, but could not ascertain what other species were caught.

We returned to *Argosy* and departed at about 1330, heading for deeper water to make a trawl haul off the mouth of Tumaco's estuaries. Two successive trawl hauls at 12 and 30 m produced a dozen small sea basses (*Diplectrum*), stargazers, at least two species of puffers, scorpionfish, a deepwater sea

Figure 48. Sightseeing in Tumaco (Vic Shifreen).

robin, and several species of flounders, including *Bothus* and *Paralichthys*. The haul also yielded several hundred brightly colored arrow crabs (*Stenorhynchus*) which promptly scurried off to all parts of the ship. Hermit and xanthid crabs, borers (*Teredo*), well entrenched in much of the decayed wood, octopods, and several species of pelecypods and gastropods completed our haul. A third tow, at 55 m, yielded seven or eight opisthognathid jawfish.

We pushed on in heavy seas toward Esmeraldas, where we would make our first official contact with the government of Ecuador. The long day had worn us out, and we turned in early.

25 SEPTEMBER 1961

We saw the lights of the city of Esmeraldas as we moved at 0415 off the estuary of Río Esmeraldas. As daylight arrived, small dugouts scooted along the coasts, some with sail, some pushed by a native laboriously sculling his craft. After a heavy breakfast, we moved inshore on *Sea Quest* but had to wait for customs clearance. To make the most of the delay, we used small hooks and still-fished in the harbor and caught two species of leatherjacket (*Oligoplites*), several jacks (*Caranx caballus*), gerreids, and small sea basses (*Diplectrum* and *Epinephelus*). Finally the customs officer arrived with our landing permit, and a letter from Bob Ellis, the fisheries representative of the Food and Agricultural Organization of the United Nations. He was attached to the Instituto Nacional de Pesca in Guayaquil, but had been recently stationed at the Inter-American Tropical Tuna Commission post in Manta. Bob's letter informed us that there had been a slight outbreak of bubonic plague (130 cases) in Manta, Ecuador, our next port of call, and that we should proceed with caution. Earlier in Miami, the U.S. Public Health Service had recommended that our party get bubonic plague shots before we left the States, but our individual doctors had considered the disease too rare for us to bother with shots.

After clearing customs, the Navy and harbor patrol checked us out, and we went ashore in a native dugout powered by an antique outboard motor. Exposed by low tide, the sand flats seemed endless. A few fishermen were laboring with their cast nets, seines, or push nets in the pools between the exposed isolated islands, competing for the fish with the ubiquitous pelicans.

The barrenness of Esmeraldas is in sharp contrast to the rich foliage of the Chocó coastline. Its waterfront is sprinkled with coconut palms and mangroves, but elsewhere only scrub vegetation dots the tablelands. Beautiful wind-sculptured beds of sandstone could be seen along the western part of the town, and we learned that these were extensive Quaternary fossil beds. The town itself is much better developed than Tumaco, with a two-lane paved highway and many buildings, but it seems to be a ghost town, with many lonely, vacant buildings.

The people were most cordial to us. In the fish market we brought specimens of everything, including a good variety of sciaenids, snooks (*Centropomus*), jacks, and barracudas. Bonefish (*Albula vulpes*) were all over, but we did not find any longfin bonefish (*Albula nemoptera*). We did smell and see a number of split and dried squid, but no fresh ones were found. The head of a large hammerhead shark (*Sphyrna*) and the saw of a sawfish (*Pristis*), both of which are reported to enter Río Esmeraldas, were in the market.

Eventually we found a curio shop run by Anton Fugazy, a fascinating gentleman whose past seemed about as colorfully nebulous as the antiques he sold. Incan artifacts were supposedly hard to come by, but we picked up some at his shop, including clay and granite figurines, toys, dogs, beads, and cups, and two carved granite monkeys. Most were Manabí artifacts of the coastal provinces, but some had been brought in from the cordilleras, and perhaps from much farther north.

On our way back to the ship, we passed the docks where larger fishes are brought in, cleaned, and quartered for market. A black marlin (*Makaira indica*) estimated at 200 kg was being chopped up, and Mr. Glassell quickly spotted four more being brought in via dugout canoe. Also being cleaned were three blue marlin (*Makaira nigricans*) from about 130 to 180 kg, four small yellowfin tuna (4 to 5 kg), two black skipjack, three skipjack tuna of about 5 kg each, and three wahoo (*Acanthocybium solanderi*) of about 7 to 9 kg. Pelagic fish are taken from dugout canoes in the blue water off Esmeraldas by hand-lining with dyed line. Some boats have cloth or plastic sails, while other fishermen merely drift-fish using bonito (for billfish) or anchovy (for tunas). When a big fish is hooked, it is played by letting it tow the boat until the fish is exhausted. Then the fish is brought alongside the boat, the boat filled with water, the fish is floated into the boat, and the boat bailed out and refloated. Quite a brave operation in these shark-filled waters.

As we returned to *Argosy*, we waved to some fishermen aboard the several Ecuadorian shrimp boats and they proudly displayed some shrimp (*Trachypenaeus* sp.) which appeared to be 12 per kg and which had been taken just offshore. While we watched, one fisherman aboard the shrimper caught a pompano similar to our Atlantic *Trachinotus carolinus*, which probably weighed at least 2 kg.

After we returned to *Argosy* we were joined by our customs agent and three armed guards whom he placed on board until we left at 1800.

26 SEPTEMBER 1961

Since 2100 last night, it has been cold enough to wear a sweater, and it seems unbelievable that we crossed the equator at daybreak. Some time during the night we had left behind water temperatures of 28.5° and 29°C, and now it was just over 23°C. The air was 18°C, but the dampness made the air feel much colder. By late morning the air temperature had warmed to 22°C.

We had no crossing ceremonies but merely tried to keep warm. A 1-m plankton tow was made over a depth of 82 m north of Manta in Bahía de Caráquez (*Argosy* Collection No. P–10) which was rich in copepods, siphonophores, and schizopods. We could see the bleak, low-lying coast of the ancient Manabí province and the bubonic port of Manta as we approached in midafternoon. Later, via radiotelephone we contacted the tuna cannery (INEPACA) owned by Van Camp. By 1615 the port captain and customs agent had come aboard and cleared us for entry into Manta. They assured us that there was no such thing as bubonic plague in Manta.

We had anchored in the harbor about 4 km offshore and east of the breakwater. The wind had subsided, so we night-lighted. The moon was full, though obscured by haze, and only a few fish were active around the light. A number of anchovies, halfbeaks, and, of all things, clingfish (Gobiesocidae) were dipped up. The schools of clupeids that came in just under the light, apparently to grab anchovies, were more elusive. We simply could not get to them. Finally, a plastic bag of rotenone was hung under the light, and, as soon as the clupeids reappeared, the bag was shot full of holes with the BB gun. The trick worked, and the rotenone seeped out into the school of clupeids, which were affected in about five minutes. About eighty were picked up, and proved to be threadfin shad, *Opisthonema libertate*.

27 SEPTEMBER 1961

The sun finally came out today and dried out our dampened bones. We were greeted by the W. R. Grace representatives, Señor Carrera, and his son Carlos. They were laden with an assortment of trinkets, Incan artifacts, some perhaps only a few days old, statues, heads, and Panama hats from Jipijapa

Figure 49. In the harbor of Manta, Ecuador, we are serenaded by a friendly native (Vic Shifreen).

and Montecristi. They wanted to trade any of our clothes for their souvenirs, so Vic promptly bartered a fine Jivaro bow and arrow set in exchange for two dirty shirts, a pair of undershorts, and one U.S. dollar. For a carton of cigarettes I got a wood carving of Don Quixote, which I subsequently found could be bought in the States for two or three dollars.

Today we were supposed to go ashore to pick up Dr. F. G. Walton Smith (Fig. 49), the director of our institute, and Dr. Richard A. Wade and Dr. Frederick B. Emerson, also from the institute, who had arrived in Manta via Aerovías Ecuatorianas from Quito. It was impossible, however, to get volunteers from *Argosy* to go into the supposedly plague-ridden city, so I went alone, quietly and quickly, to meet our newcomers at the airport. On the way back we shopped for food and supplies and stopped briefly at the INEPACA tuna cannery. On one loading ramp was an area of perhaps 10 by 30 m which was stacked to a depth of 1 to 2 m with skipjack (*Katsuwonus*) averaging 5 kg each.

We returned to *Argosy* and left for Isla de la Plata, our ultimate destination for collecting. Today was the first good breeze we have had in a while, and we took advantage of the force-4 wind to hoist sails. After dark, we made a tow about 40 km north of La Plata, with a homemade 5-m mid-water trawl with nearly 270 m of wire out. In a forty-five minute tow we got hatchetfish, several eel larvae, two species of caridean shrimp, ctenophores, and the larvae of cephalopods, decapods, and stomatopods. A 1-m surface plankton tow made simultaneously yielded a few larval fishes, some larval shrimp, copepods, and a million pieces of shredded Kleenex which virtually ruined the plankton.

28 SEPTEMBER 1961

A great brown lump is perhaps the quickest way to describe Isla de la Plata (Fig. 50). When we anchored at 0545, the overcast, cold weather did

Figure 50. The northwest coast of Isla de la Plata, Ecuador (Vic Shifreen).

Figure 51. Looking south along the eastern side of La Plata (F. G. Walton Smith).

nothing to enhance the island's dreary appearance. The island lies west of the rain line and is covered with a carpet of grasses, tree cactus, Acacias, scrub vegetation, and vines, and is virtually without large trees. Although about the same size as Gorgona, it is extremely dry, and only during the rainy season does much greenery become evident. About 5.5 km long from north to south and 2 km wide, the island has three distinct beach levels (tablazos) which reflect the thousands of years of erosion that formed them. These three levels represent successive erosion as the island underwent uplifting. A deep gorge, eroded to a depth of perhaps 150 m, is believed to be a remnant of an ancient river valley when La Plata was connected to the continental mainland.

Our anchorage was in 55 m, just off the sandy beach covered with the houses of the fishermen who barely existed there (Fig. 51). Most boats had already put out for a day's fishing (Figs. 52, 53), but as we approached the shore, we were greeted by curious children (Fig. 54) whose shyness was quickly overcome when they found we were trying to catch fish. The flat, low tide pools were heavily eroded by the surf. Along shore, sandstones and limestones were liberally pocked with sills and dikes of quartz and other magmatic intrusions, while the pools themselves were largely eroded basalt (Fig. 55). Anemones, hermit crabs, chitons, and gastropods were common in the smaller pools, where multitudes of red or blue porcellanid crabs scurried about, tenaciously holding onto a rock in spite of crashing surf.

Figure 52. Looking north along the eastern side of La Plata from *Argosy*'s anchorage (F. G. Walton Smith).

Figure 53. The crew of a bongo boat prepares to haul their beach seine at La Plata (F. G. Walton Smith).

Figure 54. Our hosts at La Plata (F. G. Walton Smith).

Figure 55. Our tidepool collecting at La Plata is greatly assisted by volunteer help (F. G. Walton Smith).

Figure 56. Dr. F. G. Walton Smith, Director of the Institute of Marine Science (Vic Shifreen).

Feeling like great white hunters, complete with pith helmets, BB gun, and camera, Clyde, Dick, Fred, and I explored the island. We were immediately greeted by one of the elder statesmen of the island who wanted to know if we had any mine detectors. When we informed him that we were looking for animals, he was disappointed, for he thought we were treasure hunters. The Manabí Incas of the mainland had used La Plata as sort of a secluded Miami Beach, and, following a weekend of revelry, they had buried caches of gold masks and figurines to appease the gods for their amoral adventures. Later Sir Francis Drake had used this island as a lair from which to dash out and plunder Old World merchant ships coming up from Cape Horn and carrying gold and silver to the countries to the north. Drake politely polished off the ships and returned with his booty, some of which reportedly is still buried on the island, hence the name "Isla de la Plata"—island of silver. In the ensuing weeks, we kept a sharp eye out while diving or trawling, but didn't see one piece of eight.

Figure 57. Señor Pedro Lucas, Mr. Glassell, and Dr. Smith in front of the angling club at La Plata (Vic Shifreen).

The only car on the island is a jeep belonging to Señor Pedro Lucas, unofficial mayor of La Plata and agent for the late Señor Emilio Estrada of Guayaquil, who maintained a small fishing lodge on La Plata for his angling associates (Fig. 57, 58). We followed the jeep tracks across the island and looked for birds, snakes, lizards, and crabs. Goat droppings were commonplace, but we never did see the former owners. Small warblerlike birds flitted in and out of the Acacias, and we could hear a variety of birdsongs in the distant foliage, but could not get a glimpse of the birds themselves. Mocking-

Figure 58. Looking westward from *Argosy*'s anchorage off La Plata (Vic Shifreen).

Figure 59. A bluefaced booby warms her young atop La Plata (Vic Shifreen).

birds were common and extremely tame, and a number were courting freely and showing no fear of us. Frigate birds perched along the eastern scarp, as common as pigeons on city hall, and occasionally one would leave its aerie to pounce on a fish below. Bluefaced (Fig. 59) and masked boobies were nesting in the long grass and soaring, particularly about the eastern edge of the island, and occasionally a tropic bird would swoop by. The cliffs on the western side of the island offer a frighteningly steep drop, at an angle of about 70°, down to the pounding, swirling surf 150 m below. Huge chunks of algae-

Figure 60. Mr. Glassell with a fine sailfish and three striped marlin taken west of La Plata (Vic Shifreen).

covered land and boulders lay scattered about like broken china. The thick talus below was evidence that the precipitous slope was eroding. The steepness of the slope and the precariously hanging boulders deterred us from further collecting in that area. Several birds and lizards plus a pailful of fossiliferous sediments were taken back to the ship. That afternoon we poisoned tide pools and obtained good results (*Argosy* Collection No. 44), despite the incessant ravages of the opportunistic frigate birds on our collection of dead and dying fishes.

Figure 61. Vic Shifreen with a broomtail grouper (*Mycteroperca xenarcha*) taken north of La Plata (Alfred C. Glassell, Jr).

By late that afternoon Mr. Glassell had returned aboard *Sea Quest* with sailfish, striped marlin (Fig. 60), and a 25-kg broomtail grouper, *Mycteroperca xenarcha*, characterized by the produced caudal filaments (see Fig. 61). After dinner, night-lighting was done from *Argosy*'s anchorage, and we were immediately greeted by clouds of krill (Euphausiacea) which so darkened the light that we could hardly see to net. Invertebrates and larval fishes were common, and we were besieged with colonial salps, each about 1 cm across, which collectively formed beautiful necklaces 15 to 60 cm long. Clyde dipped a number of squid, including one on which he had been working for his thesis, so of course he was overjoyed. In spite of intensive angling by the crew, no fish were caught, and we turned in.

29 SEPTEMBER 1961

Another windy, cold, and overcast morning was too much for us, so we worked on our equipment and slept. In the afternoon we towed the 5-m flat trawl east of La Plata but promptly lost it on a rock outcropping over what showed on the chart as an otherwise smooth bottom. Bridles, doors, and floatline were recovered, and a second trawl was rigged and fished, this time returning successfully (*Argosy* Collection No. 46), although the catch was not as good as we had hoped for. A subsequent tow 11 km off Punta Canoa gave us an excellent haul of several flounders (*Paralichthys, Bothus, Citharichthys*), sea robins, scorpionfish, sea basses, and a diversity of invertebrates including hermit crabs, brittle stars and sea stars, octopods, chitons, gastropods, and tectibranchs. A 40-cm flounder, about to be preserved in formalin, brought tears to Captain Rosing's eyes, and because we had other, smaller ones, we donated it to him and cooked it for his supper. Just like his North Sea plaice, he admitted later.

On our return to La Plata, we purchased specimens from the fishermen and also traded hooks, lines, and leaders for fish. Later that evening we dipnetted many fishes amidst the clouds of krill which glowed blue in the dip net. Strings of polychaetes which resembled the South Pacific palolo worm seemed to be everywhere. The one remaining pyramidal trap was baited with dolphin meat and a flashlight put in a quart jar. As it was lowered, at about 30 m the plastic cap imploded, but the trap seemed to be in a good location so we let it stay.

30 SEPTEMBER 1961

The pyramidal net had fished well, nevertheless, and we got a number of gobies, including a beautiful orange species with electric blue stripes, cardinalfish, postlarval scorpionfish, and several crustaceans. Earlier that morning, Red Stuart had spotted a pod of killer whales near *Argosy*, the largest of which was a male about 10 m long, accompanied by several females of about 5 m. We went out on *Sea Quest* to photograph the whales, and also to catch some bonito for bait. Along the northern, northeastern, and southeastern edges of the island we passed by dense schools of anchovies which were being herded by bonito. Before today the crew had been catching skipjack tuna, but now all we could find was the Pacific bonito, *Sarda chiliensis*, a species typical of cooler waters. Before the trip was over, we were to notice rather sudden slight shifts in the pelagic fish fauna and also in the plankton. We could also detect fluctuations in water temperature during our daily collecting dives, particularly when deep diving. The possible reason for these changes was apparently due to a tongue of cool water entering from north of La Plata.*

Later that morning Red Hagen caught a 9-kg wahoo which we welcomed because of the food shortage, and our diving had increased our appetites to where the cook could not cope with us. But we soon tired of eating wahoo daily, in spite of the seasonings we used to make its dry flesh more palatable. An 18-kg broomtail grouper was also quickly fileted and put in the larder. All around us were small tuna fishing boats, some manned by Japanese. They were jackpole fishing using lookdowns (*Selene*) for bait. Perhaps the silvery sides were good lures and other bait was being used as well, but we could not tell from this distance. On our way in we encountered several huge, compact schools of the chub *Sectator ocyurus* which we had thought were so rare, but which had darkened our night-lights earlier in Piñas Bay.

In spite of the overcast weather, the water was relatively clear, and we made a collection along the reef north of the anchorage using poison and multipronged spears. After two hours in the water, we were relieved not to have encountered any of the packs of sharks reported by anglers off La Plata. Perhaps the cooler water (22°C) had deterred them. Morays, wrasses, butterflyfish, demoiselles, blennies, gobies, and mullets were taken, and a round stingray (*Urotrygon*) was speared and two others seen. Large schools of mullet resembling *Chaenomugil* were avidly browsing on the rocks in the subtidal area, but none could be speared.

Toward the end of our dive as we were starting to swim back to the dory, I saw what I thought was a high, triangular dorsal fin. My already chilled body froze further as I realized it was a killer whale. I was still too far from the boat to swim rapidly to it, so I dived to the bottom signaling to my diving buddy, Dick Wade, to get down with me. And then two female killer whales, their outlines barely visible through the silty water, lazily swam by 10 m away. From my stance on the bottom at 6 m, I could clearly see their distinctive white neck and side patches. We quickly surfaced, but Dick had not seen the whales, had no idea of what was out there, and had not heard that they were in the area that day.

We changed into dry clothes and put out again in the dory to investigate a report by Captain Rosing of a school of manta rays doing somersaults. This sounded like a tall tale, but just off the easternmost tip of La Plata was indeed

---

* This is discussed further under Oceanographic Observations.

a school of manta rays, either *Manta hamiltoni* or *M. lucasana*, doing repeated somersaults. Dick photographed the action with the telephoto lens of his movie camera, but subsequent examination of the movies did not reveal if young were being born, which is the alleged reason for this somersaulting.

Our ritual of night-lighting continued after dinner, this time west of the island along the 90-m curve near the continental shelf, where it drops off into very deep water. Results were meager, so we towed the mid-water trawl and plankton net and got a good collection of fishes and invertebrates, including two liters of plankton (*Argosy* Collection No. P–12). At the end of the station, farther offshore, we again hove to and tried our hand at night-lighting. A 5-cm dolphin and a flying halfbeak (*Euleptorhamphus*) were taken, but the waters were poor in life. A small siphonophore containing a freshly ingested mullet was scooped up, but that was about all we saw until we heard Clyde scream with delight that a school of large squid, apparently *Dosidicus gigas* and each perhaps 0.5 m long, had come into the light. But they were wary and would not approach the net. Finally, Clyde managed to get one in the net, but it popped back out and was lost.

After several hours of drifting in the inky blackness, the night was now calm, quiet, and eerie. One's imagination became very fertile, recalling sailors' tales of giant squids reaching onto ships' decks and furtively plucking men from their watch into the black depths. Dr. Anton Bruun had recently further promoted the existence of real sea serpents, adding that they probably did live in parts of unexplored sea. He specifically mentioned waters near the Ecuadorian coast, right where we were drifting. Maybe we would find, or be found by, some strange creature. But our reverie was abruptly broken when Clyde, who was handling a small squid he had just dipped, split the silence with a choice expletive as the squid bit him.

1 OCTOBER 1961

Carlos Carrera, soldier of fortune and purveyor of one thousand and one treasures, arrived from Manta this morning and greeted us at La Plata with our badly needed groceries, supplies, beer, and an outstretched hand. After proper payment, Dick was told that the rest of his diving gear was tied up by customs agents in Manta and that we would have to retrieve it ourselves. In transit, in spite of a strong wind and a heavy following sea, we tried to tow the 1-m plankton net, but the bridle snapped. A second net was fished, simultaneously with a 5-m trawl. The plankton tow, one of the few we made in daylight, was quite poor, although the trawl catch was excellent, and Captain Rosing even got another flounder for dinner. Several species of flat-

Figure 62. Tuna clippers in Manta Harbor (F. G. Walton Smith).

Figure 63. Ashore for supplies at Manta (F. G. Walton Smith).

fish, sea basses, puffers, morays, jawfish, and shrimps, crabs, brittle stars, sea stars, bivalves, octopods, gastropods, polychaetes, and mysids comprised this living smörgåsbord. A second trawl haul apparently fished on bottom only part of the time, but we did get some specimens, including a strange portunid crab with a supplementary spine on the cheliped.

We anchored in Manta's harbor (Fig. 62) and went ashore (Fig. 63, 64) to purchase more supplies. That night we tried to night-light, but the water was extremely silty, although many balao (Hemiramphidae) were seen, and one was captured. The captain of a nearby tuna clipper, the *Francis María*, invited us to come aboard, and we shared its hospitality and freshwater showers.

Figure 64. Water casks and burros in Manta (F. G. Walton Smith).

2 OCTOBER 1961

Bob Carpenter, fleet manager for INEPACA in Manta, cordially welcomed us to visit his tuna plant. He too had noticed the extreme fluctuations in water temperatures on the tuna grounds. Three weeks ago the temperature at Salinas, where the boats were pole fishing to the south, was 18°C, and fishing had dropped off. The best fishing, Bob said, is from 23 to 24°C. The purse seiners supplying INEPACA are equipped with nets about 900 m long and 450 m deep. When the cold water extends above the leadline, the tuna are easily captured because they will not dive below this line into colder water. He believes that this is a mixing area between the Perú Current from the south and El Niño from the north, but admitted that although they did maintain records of catch and temperature, they really had very little comprehensive oceanographic data to rely on in their fishery.

Bob told us that the best marlin fishing was off Cabo San Lorenzo, La Plata, Porto López, Santa Rosa, Salinas, and Palos Verdes, facts which we promptly relayed to Mr. Glassell for future use. INEPACA employs several hundred persons who pack skipjack for Van Camp. The belly strips are a real delicacy in South America, and are packed in soy oil and sold as "Tipo de Sardinas." The boats are all for pole fishing, but the fishermen have had trouble getting bait because the Ecuadorian government had recently outlawed the practice of catching bait.

We had no luck buying specimens at the Manta fish market. Most fish in the market were the larger tuna and billfish, and their abundance leaves little market for the smaller, inshore fishes which we were seeking. We brought some ship stores and headed back, stopping in a religious shop on the way so that Dr. Smith could purchase a beautiful wooden replica of a priest, whom we later named "the Bishop of Manta." This icon would become our much-admired guardian saint and savior of the sea (Fig. 78).

Dick Wade had by now successfully pried his diving gear away from the customs agents. Before we left we met with Roger Cairns, a graduate of the University of British Columbia and now with the Inter-American Tropical Tuna Commission. The commission's office in Manta had been closed three weeks earlier, and he was between assignments. He told us that although the outbreak of the plague was over, quite a few people had been ill, and three had died.

On our way back to La Plata we ran a hydrographic transect from Cabo San Lorenzo to a point west of La Plata, taking bathythermograph readings and Nansen bottle casts from the trawl winch and snapper fishing reel. It was primitive surveying, but it worked. *Sea Quest* had stayed to fish off La Plata during our trip to Manta, but had seen few fish. The water was still cold and dirty, yet our sighting of a few sea snakes, lovers of warmer water, augured that warm water might be back.

3 OCTOBER 1961

Today was another bleak, overcast day. A solid blanket of low-lying clouds covered the ocean as far as we could see, and there seemed to be no patch of blue anywhere. The northeastern tip of La Plata was our diving and collecting area, but this proved to be virtually without cover for fishes. A scoured-out bottom with a few rocks and boulders was about all we could find, with no coral growing anywhere. Except for large schools of *Abudefduf* and *Acanthurus*, reef fish were scarce. Some parrotfish and a few black-and-orange triggerfish (*Balistes verrucosus?*) were seen but could not be approached within spear range. Moving southward along shore, we spotted a few alcyonarians, and then, just north of the eastern tip, we discovered more protected areas less subject to waves. Here the talus increased, and coral stacks were everywhere. At depths of 5 to 7 m we used two liters of rotenone, a jar of quinaldine for collecting heterocongrids, and spears with five prongs to collect the small fishes not affected by the poison. Large schools of *Sector ocyurus*, *Acanthurus triostegus*, and *Haemulon maculicauda* swam about us but were unaffected by the poison, possibly due in part to the cool water (23°C). I obtained one hour and fifty minutes bottom time from a double air tank, and our scuba gear worked well for a change, although I found that a face mask over a mustache is awkward. The time we spent was worth

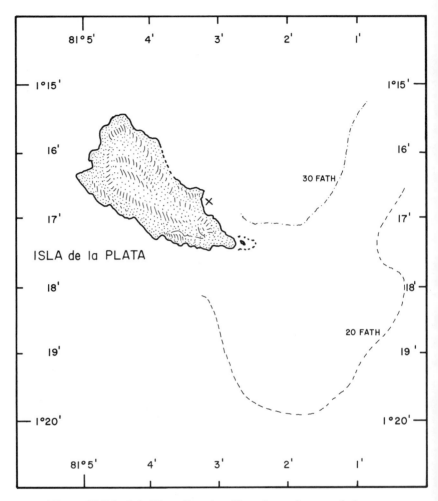

Figure 65. Isla de la Plata, Ecuador. X marks anchorage of *Argosy*.

it, and we got pomacentrids, serranids, gobies, blennies, and many other fishes (*Argosy* Collection No. 58). A few invertebrates were taken, but suitable crevices were rare, and considerable coarse silt covered the bottom, which may have made it unfavorable for infauna. A second collection, using snorkels, was made from the shore out to a depth of 5 m, and yielded essentially the same species as offshore but in much fewer numbers. A third poison station was made farther south. Many larger fish were seen, but stayed out of our spear range. They were surprisingly wary in an area where few if any divers had ventured. A large species of *Epinephelus*, with two white spots on the caudal peduncle, was not identifiable to species. Sicklefin grouper (*Dermatolepis*), surgeonfish, butterflyfish, and parrotfish swam quickly in and out of our area but stayed clear of the poison.

When we returned to the boat, Mr. Glassell had already brought back a beautiful 52-kg sailfish which he took west of La Plata. Its ovaries showed no indication of recent spawning.

We then bought some specimens which the natives had taken in a huge beach seine and on hook and line, including two nice morays and a large wrasse (*Bodianus diplotaenius*). Later we resorted to night-lighting and obtained one of our best collections to date (*Argosy* Collection No. 63). The crew fished on the bottom (37 m) and caught barracudas, snappers, jacks, a congrid eel, and a large guitarfish (*Rhinobatos planiceps?*). After night-lighting, we hauled anchor and departed for Salinas, about 110 km south on the mainland coast.

4 OCTOBER 1961

Salinas is a quiet, pretty little town just inside La Puntilla, the entrance point to Bahía de Santa Elena. This resort area swells in size in January when the more affluent throng to its beaches for recreation. We cleared with the port captain shortly after our arrival at 0730 and went on to La Libertad, home of Anglo-Ecuadorian Oilfields, Ltd., where we made arrangements for fuel, water, gas, and food. Again we had trouble getting sufficient food, and went back to Salinas where we fared no better. Eventually we were forced to return to La Libertad's marketplace, where, by going in each store, we bought up an adequate supply of fruits, vegetables, and tinned sausage to sustain the expedition members. Canned groceries were costly and almost unobtainable, but not all services were expensive. We ran into Captain Rosing, who had paid fifty cents for a haircut, a tip, and two cold beers.

Earlier we had seen tuna clippers and some small boats laden with fish moving into the harbor, so we headed for the fish market, which by now was well supplied with fresh fish. Eager to replenish our food supplies, we bought fish as long as our money held out. Large sierra mackerel, skipjack tunas, blue jacks, red snappers, moonfish, tilefish (*Caulolatilus*), parrotfish, grunts, spadefish, herring, and weakfish were packed into our burlap bags. Some fresh squid were supposed to come in that afternoon, but never did, and we went back to the ship. On our way out of the marketplace we noticed nice appetizing chunks of fresh, red meat which we nearly purchased until we found out that it was donkey.

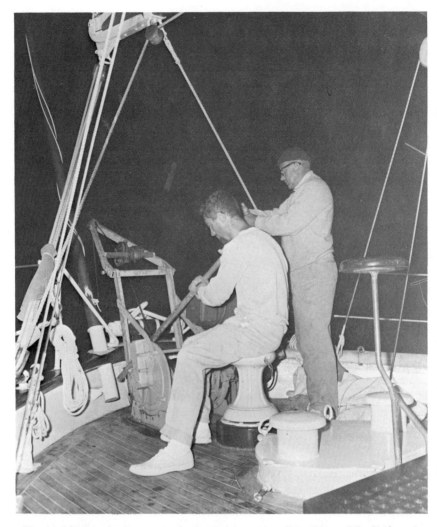

Figure 66. A bathythermograph cast being made off La Plata (Vic Shifreen).

5 OCTOBER 1961

After a rather poor evening of night-lighting in La Libertad's harbor, we headed for La Plata. During the night while at anchor, someone had made off with our double air tank, in spite of a two man watch. We continued our hydrographic transects begun earlier in the week off Manta, and made bathythermograph and Nansen bottle casts in route (Fig. 66). As usual the weather was overcast, and a strong west-southwest wind and current were driving us inshore. Southwest of Salango Island we spotted a Ridley turtle, and later a sandpiper flew aboard, but the area otherwise seemed quite devoid of life.

Dr. Smith had planned to work on his book on this trip, but had spent most of his time helping us in every way. Today he became so engrossed in the work that he spent most of his time tirelessly reading bathythermograph slides and avidly analyzing them.

Our last plankton tow for the trip, made east of La Plata, caught almost nothing. Similarly, a forty-minute bottom trawl yielded a stomatopod larva, several megalops crab larvae, and a postlarval fish. The entire collection fitted into a small vial. We arrived at La Plata after dinner and promptly fell asleep. But when Red Stuart yelled to us that the water was alive with life, all popped out of bed and started scooping. This was a superb catch, and after five hours of night-lighting (*Argosy* Collection No. 67), we had obtained one of our best assortments of larval and juvenile fishes and many invertebrates, including at least five species of Clyde's much-desired cephalopods.

6 OCTOBER 1961

We slept late today in preparation for a five-hour diving trip off the northeastern part of La Plata. Rotenone and spears were used liberally; in spite of poor visibility, fifty species of fish were collected or seen (Fig. 67), and an equal number of invertebrate species was taken. Today I saw my first live shark in Ecuador, a 2-m nurse shark (*Ginglymostoma*) which was prowling

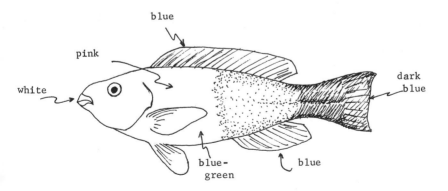

Figure 67. Parrotfish seen at *Argosy* Collection station 68 at La Plata (Robin Ingle).

Figure 68. Mako shark caught by Mr. Glassell off La Plata (Vic Shifreen).

around a cave. From a distance of perhaps 2 m, I let go of the Hawaiian sling with full force, but the thin spear bent on impact, and the shark shot away, its hide apparently undented.

Mr. Glassell caught a female mako shark (*Isurus oxyrinchus*) estimated at 135 kg (Figs. 68–70), the jaws of which were saved for our museum. Very

Figure 69. Mako shark under discussion (Vic Shifreen).

few other fish had been seen offshore in the angling grounds, and things seemed to be getting no better. Night-lighting after supper yielded the usual assortment of fishes and invertebrates, including five specimens of *Octopus* sp. with extremely long arms which gave the animal a squidlike appearance.

Figure 70. Small tunas make good bait for mako (Vic Shifreen).

7 OCTOBER 1961

After ten days at La Plata we saw the sun today, so we decided to make a dive in deeper water in hopes of better visibility and a chance to find out what was down there. None of us had ever been below 30 m, and the thought of diving off *Argosy* to nearly 40 m was met with some trepidation. In the morning we dove along the slope from 10 to 20 m, and Dr. Smith accompanied our boat to photograph our surface preparations. At depths of 14, 18, and 29 m, we encountered the fascinating garden eels (Heterocongridae) which we tried to collect using spears, poison, and anesthetic (quinaldine), but without success. We could see them by the dozens swaying with the swells and looking like a wheat field. Each time I speared one with the tiny multiprong spear, the barbless prongs would pull out as I tried to extricate the eel from its burrow. They held on with ferocious tenacity, and I was unable to get a single specimen (Fig. 71).

That afternoon we prepared for the "deep" dive from *Argosy*. We dropped down the anchor chain to 37 m, where a mysterious greenish glow seemed to pervade everything. A school of *Caranx melampygus* circled us, as several porgies (*Calamus*), each about 2 kg, lurked in the distance. A large skate (*Raja*) flipped by at 34 m, but scooted off before we could launch a spear. At 33 m we ran into a wall of cold water and a viscous layer which looked like jello. Above this refractive layer the temperature was 19°C, and just below the layer, which was perhaps 2 m thick, it was 17°C and very turbid, with a rather strong current sweeping the suspended silt northward.

On the bottom, we were greeted by an old rubber tire which housed a school of cardinalfish (Apogonidae) and a soapfish (*Rypticus*). Otherwise, the

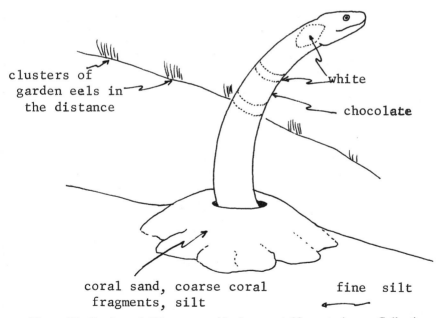

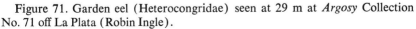

Figure 71. Garden eel (Heterocongridae) seen at 29 m at *Argosy* Collection No. 71 off La Plata (Robin Ingle).

Figure 72. Congrid eels were common at 37 m where *Argosy*'s anchor rested east of La Plata (Robin Ingle).

coarse sand bottom was virtually scoured clean of detritus, and we realized why we had been catching so little in the pyramidal net hung from *Argosy*. Nearby, we picked up a handful of hermit crabs, gastropods, and pennatulids. About the anchor, several depressions which we had hardly noticed suddenly came to life, and each revealed the head of a conger eel, perhaps 10 to 13 cm thick (Fig. 72). It was then I noticed that Clyde was starting to pick one up. As he was about to grab it, I quickly swam after him and jerked his hand away. The eel started out of its burrow, and I speared it. And out came an eel perhaps a meter long. It fought hard and corkscrewed around the spear and my wrist, but we managed to get it up into the boat, slime and all. As we shed our gear, Clyde explained that he had not had his glasses on and had thought it was a sea cucumber.

At that depth, we had only fifteen minutes bottom time with our single tanks including about five minutes for collecting. Dick's teeth hurt badly from the pressure change, but that soon passed, and we resumed working on our gear and collections.

8 OCTOBER 1961

The sky was still bright but overcast, and Dick took movies around the island. The rest of us made tide pool collections and pulled the beach seine. Some of the poisoned fish attracted the ravenous frigate birds, and they again moved in on us to compete for our valuable specimens, which probably by now were costing us $100 a pound. The hauls were not particularly rich, nor were the tide pools full of life, but we could collect intensively because the surge had subsided. That afternoon, a collecting dive was made south of *Argosy* over a rich coral park. Very large groupers (*Epinephelus analogus* and *Mycteroperca xenarcha*), brown with chainlike markings and diaphanous eyes, swam about us eating the fish as fast as we poisoned them. Soon they were joined by amberjacks (*Seriola dumerili*?) which must have weighed 50 kg. When we first saw them from afar they looked like sharks. The amberjacks swam less than a half meter from our face masks and made us very

Figure 73. Mr. Glassell and Red Hagen display one of the many roosterfish (*Nematistius pectoralis*) caught near Salango Island off mainland Ecuador (Vic Shifreen).

uncomfortable. Generally they swam in pairs, rolling their eyes in a most humorous manner. On several occasions I poked a spear gently at them, and once, when I grabbed one's caudal peduncle, I nearly had my arm yanked from the socket. But even so, they continued to mill about us incessantly. Large schools of snapper (*Lutjanus* and *Paranthias*), filefish (*Xesurus*), and grunts (*Haemulon maculicauda*) swam about us in such numbers that they

Figure 74. Broomtail grouper, roosterfish, bonitos, sierra mackerel, and jacks are numerous off Salango Island (Vic Shifreen).

obscured our vision. The light dimmed rapidly, and by 1700 we ceased diving because we could see no more than 2 or 3 m.

Today Mr. Glassell fished near Salango Island, off the mainland coast. He took five black skipjack of about a kg each to use for marlin bait, six broomtail grouper up to 10 kg, three crevalle jacks to 3 kg, three spotted jacks (*Caranx melampygus*) to 3 kg, and eight roosterfish (*Nematistius pectoralis*) up to 10 kg (Figs. 73, 74). A real good day's fishing had yielded some fine specimens for our collection.

After night fell, we took gasoline lanterns in the dory over to the inner reef north of the anchorage, where we had seen many fishes. Except for a few needlefishes (*Strongylura stolzmanni*), we saw no activity at the surface. From the side of the boat, I looked underwater with my face mask and saw a few angelfish, but the reef, which had been so teeming with life in the daytime, was now barren of fishes. After a second thought I decided not to try diving at night, and instead we fished on the bottom with hook and line, catching only one sea bass (*Paralabrax*) with white spots. Rather abruptly to the east of us, we heard considerable splashing on this quiet night, and quickly moved over to where the action was. Over a depth of about 37 m, we dipped up some of the flopping fish, which were small silversides (*Mugilops*), a few goatfish (*Mulloidichthys*), and again some *Octopus* larvae. Fish activity ceased as suddenly as it had started, and we returned to *Argosy*.

9 OCTOBER 1961

By noon the overcast sky had broken up and patches of blue were now everywhere. We dived north of the anchorage, hoping to use the Fenjohn underwater movie camera which, up to that time, had been useless in the dimly lit waters. But after 2 m of film had been shot, it failed to operate, and we were out of luck just as huge schools of multicolored fishes descended on us.

Clyde saw his first shark in Ecuador, and was quite shaken after he had poked his head into a cave where a nurse shark of about 2 m was sleeping. These seem to be different from the Atlantic form (*Ginglymostoma cirratum*) in their pattern and gray dorsal coloration. A beautiful cornetfish (Fig. 75) of about 1 m was seen but could not be speared. Although it appeared sluggish, it was deceptively fast when pursued. Our poison did not work well today, possibly because it is getting old. The few fish that we did obtain with poison became lodged in crevices, and we had to fight the morays from getting them before we did. At one point a golden-spotted green moray nearly 2 m long started to come after me, then backed into its lair, but not before it had thoroughly scared me. Shades of Gorgona Island!

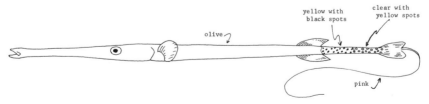

Figure 75. Cornetfish (*Fistularia*) seen at *Argosy* Collection station 78 (Robin Ingle).

Mr. Glassell returned on *Sea Quest* with two striped marlin (*Tetrapturus audax*) of about 55 kg each, a sailfish of about 50 kg, a wahoo, and two yellowfin tunas of about 14 kg. The marlin and sailfish were females, but the ovaries showed no indication of spawning. Stomach contents of all fish were preserved. Both Mr. Glassell and Red Stuart had fished these waters many times before, but they claimed that this time was the poorest fishing they had ever experienced. We were not too happy with the cold water in which we were diving, and the fish may have felt the same way.

Night-lighting tonight yielded few fishes but many squids. At least four species of cephalopods were taken, and at one time Clyde took sixteen in a single swoop of the net. Large swarms of krill and megalops larvae were again beginning to appear. We lowered the pyramidal trap net from the stern of *Argosy*, set a 25-hook trotline baited with wahoo and yellowfin tuna off the nearest reef, and strung the gill net in a cove between two coral reefs.

10 OCTOBER 1961

Our entire catch in the pyramidal trap was several cardinalfish and a post-larval scorpionfish, all in fourteen hours of fishing. The trotline fared only somewhat better, as only two of the twenty-five hooks were retrieved; the rest had just disappeared. But we did get a very pretty snapper on one of the remaining hooks (Fig. 76). Our gill net contained two large chubs, *Kyphosus analogus*, and about twenty soldierfish, *Myripristis* sp. Soldierfish were usually

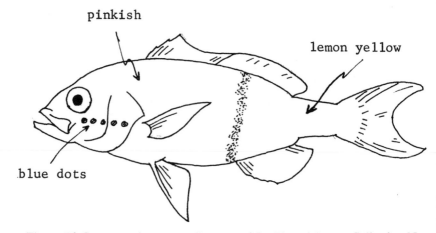

Figure 76. Snapper taken on trotline east of La Plata (*Argosy* Collection No. 82) (Robin Ingle).

seen to lurk in holes in the reef during the day, and evidently they roam freely by night far from the reef. These were caught nearly 100 m from the nearest shelter.

In the afternoon we decided to turn all our energies into capturing the garden eels along the deep slope. Dr. Smith had repaired the underwater movie camera, and we wanted to make movies of the eels' behavior. We dived down the slope to 30 m, again passing through a visible thermocline which was now between 27 and 28 m. The northward current, up to 2 or 3 knots, was scouring the bottom along the entire slope, and we had to use extra weights and hang onto anything we could grasp to keep from being swept away. It was almost impossible to make headway except by pulling ourselves along the bottom. Visibility was extremely poor above the thermocline although it was clear and icy beneath it. We finally found the eels clustered in several patches at from 21 to 26 m, and although it was too dark to get an adequate light reading, Dick took movies anyway in hopes of some good luck. I crept over a patch of eels and squirted quinaldine into several burrows.

Finally one about 15 mm thick came out, and I shot it, but the spear did not hold, and it quickly snapped back into its burrow. But by now we were exhausted from fighting the current and were low on air. Dick signaled that he had five minutes left, pulled his J-valve, and headed for shore, camera in tow. Clyde and I acknowledged and attempted to follow him when suddenly he disappeared before our eyes. We tried to follow the bottom where the current was less but could not spot him, so we surfaced just south of the anchored dory. Then we spotted him downstream of the current, gasping for breath. I swam to him, and he gave me his empty tank which he asked me to carry back to the boat, rather than use it himself as a float. With Clyde and I standing by, Dick made it on his own back into the boat, exhausted but safe. We were all pretty scared at the close call in these murky waters. He had run out of air rather suddenly, had become exhausted, and tried to squeeze his life jacket belt, but it had failed to function. Dick jettisoned twenty-four pounds of lead plus the Fenjohn camera to get to the surface.

Clyde and I spent the next hour taking turns towing each other to search for the camera. While the thought of being trolled for sharks did not really appeal to us, we did want that camera. Just as visibility was about gone, we spotted the weight belt. We anchored, and as I descended along the anchor line I spotted the camera, which was resting not more than 3 or 4 m from the weight belt. On the way up a large school of *Mulloidichthys* circled, and on the next dive I was able to spear one. We had gotten the young stages, but these were the first adult goatfish we had taken.

Then we went ashore to purchase specimens the fishermen had promised us in exchange for cable, swivels, hooks, and feathers. We got several large flounders, some snappers and grunts, an angelfish, a small dogfish with white fin-tips,* a flame crab (*Calappa*), and a large hermit crab, none of which we had seen before. But it was difficult to explain to the fishermen that we were buying these to put them in alcohol, and their cooperation began to diminish as they realized what we were going to do.

That night, before we started to night-light and set the pyramidal trap, our pet kinkajou, who had been *Sea Quest*'s mascot since Buenaventura, became nasty and bit several of the party. Tempers flared, and we knew, after nearly six weeks of cramped living together, that the cruise was beginning to affect our nerves.

11 OCTOBER 1961

The day was overcast and gloomy, the water turbid, and the wind cold. Although I still was determined to get these garden eels, no one really felt like diving, and several of the party were starting to come down with colds. Captain Rosing had been on his back for nearly a week with a severe viral infection. All in all, and with yesterday's scare, we decided to turn our efforts to packing the specimens for shipment and to cleaning up our gear. Yesterday had been our last dive of the trip, and we wanted to insure that we all got back. Dr. Smith and I went ashore to collect rock samples. Also taken were ghost and porcellanid crabs which we got with the BB gun, and we found numerous hermit crabs (*Coenobita compressa*) inhabiting the holes in the rocks on the cliff face. When we returned to *Argosy*, Dick and Clyde had pulled the pyramidal trap and had obtained several sizes of fishes not taken earlier (young stages of *Paralabrax*, *Apogon*, and *Canthigaster*).

In the late afternoon, Mr. Glassell returned with a 91-kg black marlin, the smallest, he said disgustedly, he had ever caught. This was particularly annoying since he holds the world record for black marlin. But the scientists were delighted because we had seen only a few black marlin, and even fewer female ones. Most small black marlin are males, and we badly needed the gonads of any marlin for our studies.

12 OCTOBER 1961

Argosy left La Plata at 0300 bound for the continental slope west of the island. By now the wind had freshened, and the sea was quite choppy when

---

* *Triakis acutipinna* (Galeoidea, Triakidae), a new species of shark from Ecuador, by Susumu Kato, Copeia, 1968 (2): 319–325.

Figure 77. Dr. Smith and Mr. Glassell atop La Plata (Vic Shifreen).

Argosy hove to at 0415. As soon as the night-light was put over the side, squids immediately darted to the light, swimming to the surface, reversing suddenly, and disappearing like ghosts. We hooked about four on squid jigs, but all dropped off. They are extremely clever and agile. Finally, Clyde teased a small piece of bonito on a jig past a squid, and Dick came from behind and got it with a net. It was a beautiful specimen of *Dosidicus gigas* about 50 cm long. Farther south off Peru and Chile they reach nearly 2 m, but we were all happy with our specimen. At first crack of dawn, the squids suddenly disappeared, so we headed back for La Plata, taking more bathythermograph casts and a few salinity samples en route.

By late in the morning, sickness from virus and *mal de mer* had begun to emaciate nearly all our party. Captain Rosing had not eaten in four days, and we later found when we reached Salinas that he had pneumonia. Dr. Smith, in spite of a slight touch of virus, and following a cold beer at 0630, decided to analyze bathythermograph slides. We headed south toward our destination at Salinas, arrived there early in the afternoon, and set about wrapping our collections.

13 OCTOBER 1961

Today was spent wrapping and crating and in contacting authorities for gear shipment. We needed a good rest, and we took it.

14 OCTOBER 1961

We arrived at La Libertad harbor at 0700 where we were met by a launch carrying eight shipping crates. Fish and supplies were packed, and we said our temporary goodbyes to Mr. Glassell, Dr. Smith, and Vic Shifreen, then departed for Anglo-Ecuadorian Oil Fields, Ltd., for final transferral to that most inelegant but functional Latin American vehicle, the *mixto*, for our final trip to Guayaquil. The *mixto* is half truck, half bus, and seems to hold an infinite number of people. Late that evening we arrived after a highly memorable three-hour trip of photographing the countryside and singing folk songs with the passengers. We took the six 100-liter milk cans of fish to the airport and shipped them to Miami air freight for $536.70, not a bad price for the work that had gone into collecting them.

15 OCTOBER 1961

After an overnight stay at the Hotel Humboldt International, we shopped about town where we got a haircut for forty cents and a shoeshine for four cents. Fred stayed on in Guayaquil, Dick flew to Quito, and I left on the night plane for Miami. The end of the trip had come for us. It had been a successful one, and we were most grateful to Mr. Glassell for complete support of the expedition in every way. Looking back, we believe that we met the original scientific objectives of the trip. In fact, we were able to gather much more information than we had planned on. We had collected many

Figure 78. The scientific party and crew at the end of the expedition at Salinas, Ecuador. Seated and revered between Mr. Glassell and Dr. Smith is "The Bishop of Manta," our savant and inspiration (Vic Shifreen).

animals, including a few new species previously unknown to science, and had greatly extended to known geographic range of some of the animals. In addition to meeting some fascinating people from the countries who were our hosts, we also learned to accept the bad luck that occasionally befalls every expedition. And perhaps an even more important lesson we learned was to live and work together and to respect each other's individuality.

Oceanographic Observations

Although *Argosy* was not equipped for routine oceanographic observations, we attempted to obtain a rudimentary knowledge of water characteristics based on a few temperature and salinity observations. We noticed that during our stay at Isla la Plata, sudden, rather extensive shifts occurred in the water temperature and concomitant fish schools, both in numbers and species. The extent of tidal mixing is not known, except that tongues of cool water were detected frequently by our divers during the three-week period at Isla la Plata, but since these tongues occurred sporadically rather than daily, it is believed that they are probably essentially nontidal in origin. It has been assumed that for the present discussion during the ten days elapsed between the first and last days there were negligible changes in temperature and tidal structure; that such assumptions may be unwarranted are realized by the author.

Three transects were made (Fig. 79) from the Ecuador mainland to just beyond the 180-m curve. At most of these (Table 3), bathythermograph casts were made to a maximum depth of 250 m, depending upon water depth, and at eight stations water samples were taken with a Nansen bottle for salinity determination to depths of 150 m.

The area between Cabo San Lorenzo and Punta Santa Elena is relatively shallow (Fig. 80), sloping gradually except north of Salinas at about latitude 2°S and northwest of Isla la Plata and west of Cabo San Lorenzo.

Bathythermograph casts reveal a steep thermocline at every station ranging from about 30 to 50 m deep (Fig. 81, 82). The abruptness of this was frequently felt by the divers when diving in excess of 30 m, and was also evidenced by a thin, well-defined light-refractive layer varying from a depth of 20 to 40 m, the latter being the maximum diving depth.

The upper layer is virtually homogeneous to a depth of from 10 to 20 m (Figs. 83–85). The well-defined thermocline appears at 10 to 28 m at its upper limit to 30 to 50 m at its lower limit, with a very steep gradient of

$0.7°C$ change per m at Station 2, to a more gentle gradient of about $0.2°C$ per m at Station 12. In sections A–A$_1$ and B–B$_1$ (Figs. 83, 84), there is some indication of downsloping of the thermocline closer to shore, but this is reversed in section C–C$_1$ (Fig. 85) in which some downsloping occurs off-shore. The stable thermocline is almost horizontal and shows little indication of breaking up except at Station 5, where there is some indication of upwelling (Figs. 82, 83, 85), and at Station 14, where a slight degree of instability occurs.

Below the thermocline, the water is homogeneous and shows evidence of thorough mixing. The top 10 m show little significant features (Fig. 86, 87), but at depths of 20 and 30 m (Figs. 88, 89) two lenses of cool water are seen off Isla la Plata and in Bahía de Santa Elena. The origin of these lenses is unknown; presumably the more northern lens moves in from deep water over the shelf, as indicated by the slight upwelling in Figures 82, 88, and 89. Perhaps the southerly lens is cut off through tidal action and retains some of its identity through its density.

Surface salinities were not plotted because they were virtually uniform along that part of the Ecuador coast in question (Table 3). Cross sections along transects B–B$_1$ and C–C$_1$ (Figs. 90, 91) show a well-layered situation, with no indication of sudden admixtures of fresh water entering from the Río de Esmeraldas far to the north.

It is thus assumed that the extensive changes in the numbers of fish seen, the species observed and collected, and the fluctuations in species composition and number taken at our nightlight and other stations off Isla la Plata were caused by the intrusion of cold, upwelled water northwest of Isla la Plata.

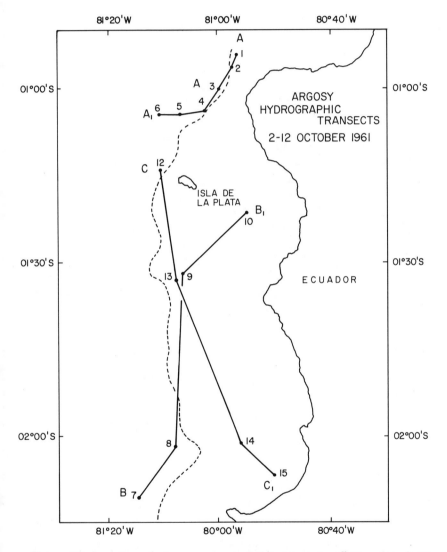

Figure 79. Location of *Argosy* oceanographic transects off Ecuador, 2–12 October, 1961. Dashed line is 180-m curve (Richard Marra).

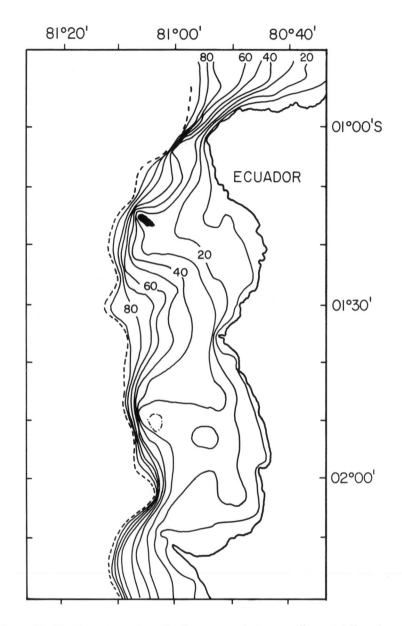

Figure 80. Depth contours in collecting areas of *Argosy* off coastal Ecuador, September–October 1961. Isobaths in fathoms. Dashed line is 180-m curve (Richard Marra).

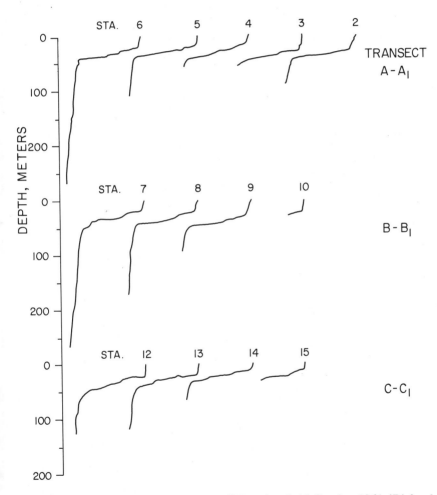

Figure 81. Bathythermograph traces off Ecuador, 2–12 October 1961 (Richard Marra).

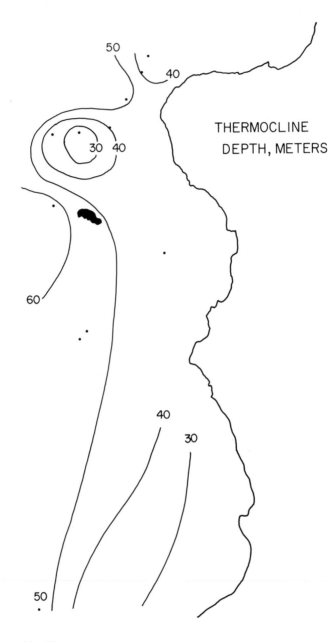

Figure 82. Thermocline depth, meters, off Ecuador, 2–12 October 1961 (Richard Marra).

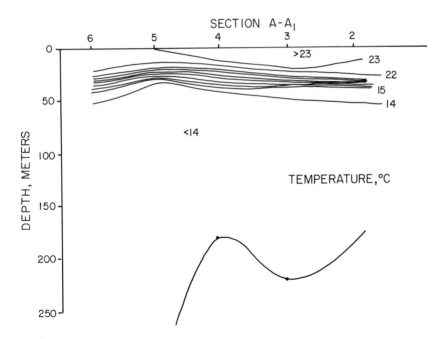

Figure 83. Temperature structure at section A–A$_1$ off Ecuador, 2–12 October 1961 (Richard Marra).

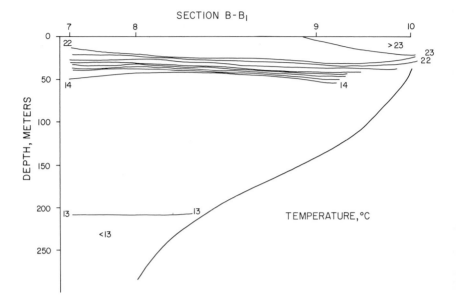

Figure 84. Temperature structure at section B–B$_1$ off Ecuador, 2–12 October 1961 (Richard Marra).

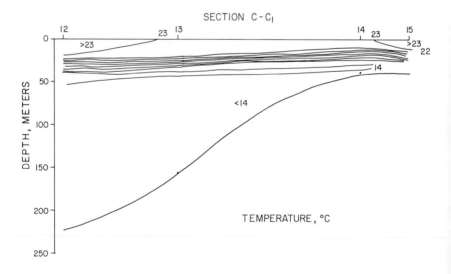

Figure 85. Temperature structure at section C–C$_1$ off Ecuador, 2–12 October 1961 (Richard Marra).

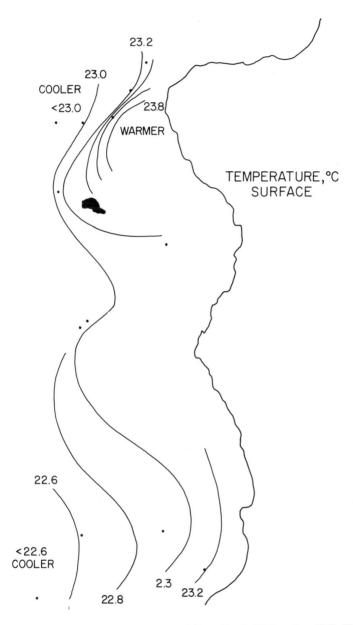

Figure 86. Surface water temperature off Ecuador, 2–12 October 1961 (Richard Marra).

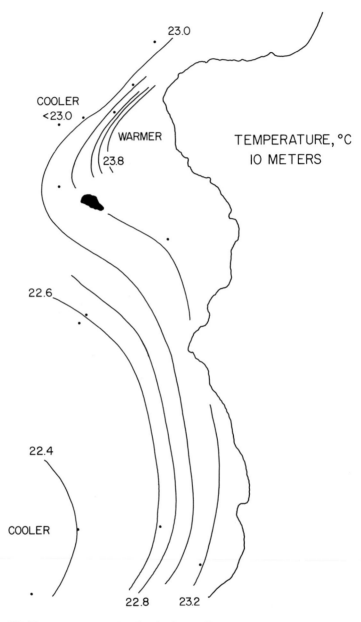

Figure 87. Temperature at depth of 10 m off Ecuador, 2–12 October 1961 (Richard Marra).

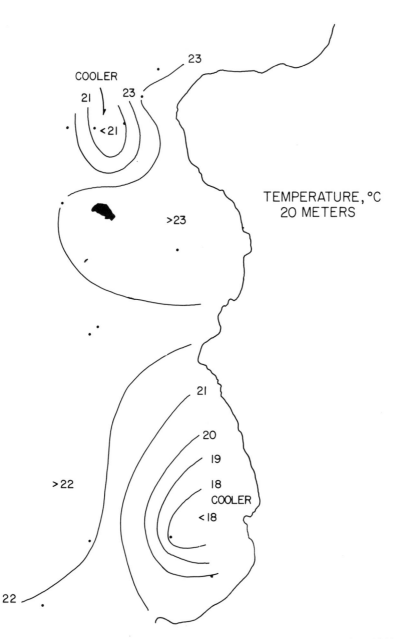

Figure 88. Temperature at depth of 20 m off Ecuador, 2–12 October 1961 (Richard Marra).

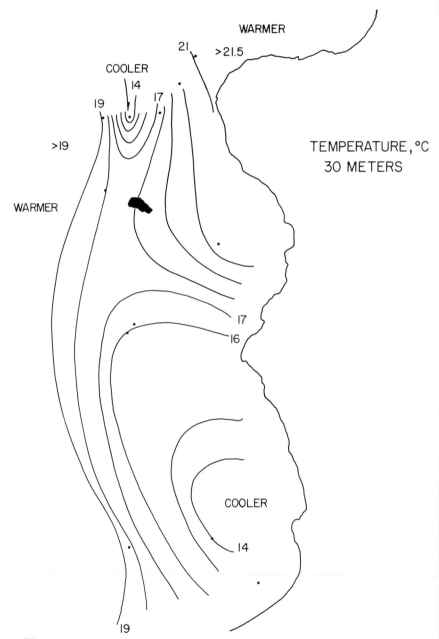

Figure 89. Temperature at depth of 30 m off Ecuador, 2–12 October 1961 (Richard Marra).

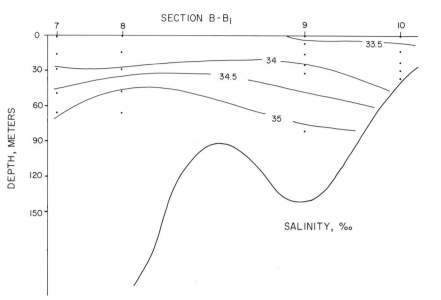

Figure 90. Salinity structure at section B–B₁ off Ecuador, 2–12 October 1961 (Richard Marra).

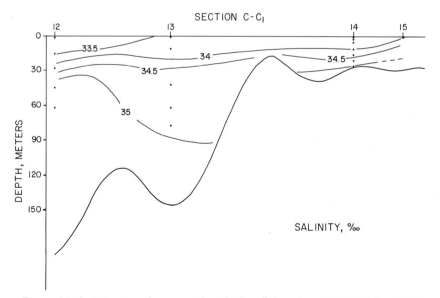

Figure 91. Salinity structure at section C–C₁ off Ecuador, 2–12 October 1961 (Richard Marra).

Appendix

TABLE 1. *Station data of Argosy collections, September–October 1961, Panama, Colombia, and Ecuador*

Sta. No.	Date	Location	Time, EST	Depth of water, meters	Depth of capture, meters	Method of capture	Bottom	Shore	Water	Tide	Wind dir. & force, Beaufort	Cloud Cover, %	Remarks
1	6 Sept.	Panama; fish market, Panama City.	—			market purchase							All specimens said to be from Panama Bay.
2	7 Sept.	Panama; 4 km E of Punta del Conchelón, Isla del Rey, Perlas Is., Panama Bay, 08°21'N, 78°46'W.	1500	30	surface	dip net			clear, green		SW 1	90	Water temperature 27.8°C.
3	7 Sept.	Panama; in surf 2 km N of Punta de Cocos, Bahía Santelmo, Isla del Rey, Archipiel. de las Perlas, Panama Bay 08°18.0'N, 78°52.5'W.	1630–1730	0–2	0–2	beach seine	black volcanic sand	black volcanic sand beach	green, turbid; 1–2 m visibility	low		90	Water temperature 27.8°C; strong surf.
4	7 Sept.	As in 3 above; in tide pools.	1630–1800	0–3	0–3	poison	bedrock, sand, rocks	lava flows; black sand beach	green, turbid; 1–2 m visibility	low		90	Water temperature 27.8°C; heavy cover of fine algae on rocks of tidepools.
5	7 Sept.	Panama; NNE of Punta de Cocos, Bahía Santelmo, Isla del Rey, Archipiel. de las Perlas, Panama Bay, 08°18.0'N, 78°52.0'W.	2100–2300	9	surface	dip net, night-light							>300 m offshore.
6	8 Sept.	Panama; Piñas Bay, in cove between Punta Molina and Punta Piñas, 07°34.5'N, 78°12.0'W.	1600–1800	0–1.5	0–1.5	poison, dipnet	sand, silt, debris	heavily wooded	dirty white; greenish	low	SW 1	100	Quiet backwater with freshwater streams nearby.
7	8 Sept.	Panama; Piñas Bay, at anchorage, 07°33.5'N, 78°12.0'W.	2000–2400	—	surface	dip net, night-light	sand	wooded	slightly turbid	ebbing	SW 1		
8	8 Sept.	Panama; Bahía Piñas, from mouth of bay to shallow end, 07°33.0'N, 78°11.5'W to 07°34.5'N, 78°12.0'W.	2030–2215	9–15	9–15	5-m balloon trynet	sand, silt, debris	wooded rain forest		ebbing	SW 1		
9	9 Sept.	Panama; Bahía Piñas, at shallow end along crescent beach, to E end of estuary at Santa Dorotea, 07°34.5'N, 78°11.5'W.	0800–1000	0–1.5	0–1.5	beach seine	sand, silt, rubble, debris	sand, rushes	brown, very turbid	slack low	SW 2	20	Heavy surf along beach; fresh and salt water.
10	9 Sept.	Panama; Bahía Piñas, near mouth of estuary at Santa Dorotea at shallow end of bay; pools along shore among igneous rocks, 07°34.5'N, 78°11.5'W.	1000–1200	0–2	0–2	poison, dip net	silt over gravel and rock, some sand	rain forest	white, clear	flood, just after slack low	SW 2	50	Air temperature 27.5°C; fresh water flow heavy.
10a	9 Sept.	As in 10 above; along sand beach.	1000–1200	0–1.5	0–1.5	poison, dip net	sandy, with some rocks	sandy cove	white, clear	flood, just after low	SW 2	50	
11	8–9 Sept.	Panama; Bahía Piñas, at anchorage, 07°33.5'N, 78°12.0'W.	2300–1300	15	15	pyramidal trap net	sand, fine mud	wooded			SW 2	90	Trap fished on bottom all night through following noon.
12	9 Sept.	Panama; Bahía Piñas, midway between E and W coves, 07°33.5'N, 78°11.5'W.	1600–1730	0–10	0–10	spear; by hand	bedrock, boulders	wooded rain forest, boulders	clear, green	slack, high	SW 1	10	Moderate to strong surge; surface temperature 27.5°C.
13	10 Sept.	Panama; Bahía Piñas, cove in west part of harbor, 07°33.5'N, 78°12.0'W.	0900–1230	10	6–10	poison, dip net	bedrock, boulders, gravel, sand	wooded, boulders	clear to slightly turbid, green	slack low	SW 1	40	Moderate to strong surge; fine, brown algae covering rocks on bottom.
14	11 Sept.	Panama; Bahía Piñas, near mouth of freshwater stream on NW side of bay, 07°34.5'N, 78°12.0'W.	1000	0–5	0–5	poison, dip net	smooth rocks and pebbles	rain forest	clear	slack low			Width of stream 2–3 m; current 1–2 kph; *Polynemus* from salt water.
15	11 Sept.	Panama; Bahía Piñas, freshwater river of E side of bay, about 2 mi upstream from Santa Dorotea.	1600–1630	0–5	0–5	poison, dip net	mud, rocks, pebbles	rushes	clear, brown				Width of stream 2–2.5 m quiet backwater.
16	12 Sept.	Panama; 60 m E of Morro de Piñas, mouth of Bahía Piñas, along sides of submerged pinnacle, 07°32.5'N, 78°13.0'W.	1000–1300	4–20	4–18	poison, spear	rock with crevices		blue oceanic below 6m; turbid above.				Some surge; visual thermocline seen at 6 m.

Sta.	Date	Locality	Time	Depth	Surface	Collection method	Substrate	Shore / bottom	Water	Tide	Wind	%	Remarks
17	13–14 Sept.	Panama; Panama Bay, W of seamount, 2000–0100, 110 km SW of Bahia Piñas, approximately 07°10′N, 79°00′W.		>1800	surface	night-light, dip net, shark hook	———	———	blue, clear	———	SW slight	0	surface temperature 27–28°C.
18	14 Sept.	Panama; Panama Bay, on seamount, 07°10′N, 79°03′W.	1900–2215	<200	surface	night-light, dip net	———	———	blue, clear	———	SW 0–1	0	surface temperature 27–28°C.
19	15 Sept.	Colombia; Bahia Solano, W side, about 6 km NW of Ciudad Mutis, in tide pools, 06°15.0′N, 77°24.5′W.	1600–1630	0–5	0–.5	poison, dip net	rocks, pebbles, sand	rocky	clear	flooding	WSW 1	80	Slight surf, medium swells.
20	15 Sept.	Colombia; Bahia Solano, W side, about 6 km NW of Ciudad Mutis, north of tide pools, 06°15.0′N, 77°24.5′W	1630–1650	0–1	0–1	poison, dip net	sand	sand beach	very turbid	flooding	WSW 1	80	Slight surf.
21	15 Sept.	Colombia; Bahia Solano, W side, freshwater stream near Collection Sta. 19, 06°15.0′N, 77°24.5′W.	1700–1730	0–5	0–5	poison, dip net	gravel, sand, rocks	wooded rain forest	clear, white	flooding	———	80	Flow 3m/sec.
22	16 Sept.	Colombia; Bahia Solano, E end of bay just off Ciudad Mutis, 06°15.0′N, 77°23.0′W.	2100–2400	75	surface	night-light, dip net	———	———	visibility about 5 m	ebbing	———	———	Anchored 180 m offshore; several species taken by angling.
23	16 Sept.	Colombia; Bahia Solano, 7 km NW of Ciudad Mutis, 30 m from shore, 06°16.5′N, 77°25.0′W	1000–1200	0–5	0–5	poison, dip net, spear	hard bedrock, boulder, rubble	rocky, sand beach	visibility 4–6 m	ebbing	———	0	Heavy swells, moderate in surge channels.
24	16 Sept.	Colombia; Bahia Solano, SE of Sta. 23 in sandy cove, 0 to 30 m offshore, 06°15.5′N, 77°24.5′W.	1300–1430	0–2	0–2	poison, dip net	sand	sand beach, rain forest	visibility about 3 m	slack low	———	0	Heavy surge.
25	16–17 Sept.	Colombia; W of Bahia Solano, about 15 km W of Punta San Francisco Solano, 06°18.5′N, 77°34.0′W.	2440–0115	750	750	5-m balloon trawl fishing mid-water	———	mountainous, rain forest	turbid, rather dirty	flooding	SW 1	0	Current moving northward.
26	17 Sept.	Colombia; NW of Buenaventura, 03°57′N, 77°35′W, about 19 km offshore, along 90-m curve, bearing 160°; two hauls	1845–1920 2125–2155	200	200	5-m balloon trawl	fine brown or black sticky mud	low, wooded	dirty, green	———	———	20	Surface water temperature 28.5°C.
27	21 Sept.	Colombia; Gorgona Is., NE tip, 03°00.5′N, 78°11.0′W.	1100–1300	10–13	10–13	poison, dip net	coral, sand, rubble, detritus	rocky, basalt and granite; rain forest	clear, blue; Secchi reading: 25 m	ebbing	———	20	*Padina* sparse on bottom; water temperature 27°C on bottom.
28	21 Sept.	Colombia; pebble shore near NE end of Gorgona Is., 02°59.0′N, 78°11.5′W.	1130–1200	0–1	0–1	poison, dip net, seine	boulders, pebbles	boulders, pebbles, rain forest	clear	slack high	———	20	Slight flow, water temperature 25–27°.
29	21 Sept.	Colombia; Gorgona Is., freshwater stream near NE end of Gorgona Is., 02°59.0′N, 78°11.5′W.	1200–1300	0–5	0–5	poison, dip net	rock, pebbles	rocks, pebbles	clear	———	———	20	Swift stream about 45° slope; some shallow pools.
30	21 Sept.	Colombia; Gorgona Is., rocky shore near NE end of Gorgona Is., 02°58.5′N, 78°11.5′W.	1500–1600	2–8	2–8	poison, spear	boulder, fine pebbles, scattered coral	boulders, basalt, granite	clear	ebbing	W 3	50	3–30 m offshore; water temperature 27°; slight surf.
31	21 Sept.	Colombia; Gorgona Is., NE shore, in tide pool 02°58.5′N, 78°11.5′W.	1600	0–2	0–2	poison	rocks, gravel	rocks, gravel	clear	low	———	———	Tide pool fresh at low water; pool 2×6 m.
32	21 Sept.	Colombia; Gorgona Is., E side, 2 km S of N tip at anchorage, 02°59.0′N, 78°11.0′W	2015–2330	25	surface	night-light and dip net; hook and line	———	wooded	blue, clouded with detritus and plankton	slack low to flood	SW 1–3	———	Slight northerly current; rainy and windy, but sea smooth.
33	22 Sept.	Colombia; Gorgona Is., E side, 2 km S of N tip, 60 m offshore, 02°59.0′N, 78°11.0′W.	0930–1015	20	20	spear	silt, coral and shell fragments	pebbles, boulders	clear, blue	high	———	100	Salinity 30.5‰; slope very steep.
34	22 Sept.	Colombia; 0.5 km NNE of Punta Mona, Gorgona Is., 02°57.5′N, 78°12.0′W.	1100–1300	5–6	5–6	poison, spear	branching coral, some coral heads	boulders, sand	clear blue	high	S 3	80	
35	22 Sept.	Colombia; Gorgona Is., E side, about 2 km S of S tip, 02°55.5′N, 78°11.0′W	1330–1400	.5–4	.5–4	spear	boulders, sand, coral	boulders	visibility about 10 m	high	S 3	80	
36	23 Sept.	Colombia; Gorgona Is., about 2 km NNE of S tip, 02°56.5′N, 78°13.5′W.	1550–1645	7–10	3–7	spear	coral (extensive patches of *Porites*), silt	rain forest, boulders	clear, slightly green	flooding	SW 1	100	Water temperature 27.8°C; current moderate, northerly.

TABLE 1—Continued

Sta. No.	Date	Location	Time, EST	Depth of water, meters	Depth of capture, meters	Method of capture	Bottom	Shore	Water	Tide	Wind dir. & force, Beaufort	Cloud Cover, %	Remarks
37	23 Sept.	Colombia; Gorgona Is., about 2 km NNE of S tip, 02°56.5'N, 78°13.5'W.	1330–1430	30	30	spear	silt, sand, detritus	wooded; boulders	clear, suspended detritus	flooding	SW 1	—	Very steep slope; slight flow.
38	23 Sept.	Colombia; 5 mi S of Gorgona Is., 13 km offshore, 02°39'N, 78°38'W.	2000–2030	100	100	5-m balloon trawl			clear, with detritus	flooding	SW 1	100	
39	23 Sept.	Colombia; SW of Gorgona Is., 20 km offshore, W of Punta Guascama, 02°39'N, 78°38'W.	1900–1945	100	surface	night-light, dip net			clear, with detritus	flooding	SW 2	80–100	Water temperature 27.0°C; night very bright.
40	24 Sept.	Colombia; about 6 km N of Tumaco, in estuary of the Río Rosario, 01°50'N, 78°45'W.	1600–1830	10–30	10–30	5-m balloon trawl	sand, muddy sand	mangrove swamp	turbid	flooding	SW 3–4	10	Water temperature 28.5°C; opisthognathids from 55 m.
41	25 Sept.	Ecuador; Esmeraldas, in harbor and fish market, 00°57.5'N, 79°42.5'W.	0945–1130	10–15	10–15	purchased; hook and line	sand, sandy mud	sand beach	muddy to clear	ebb to flooding	SW-W 3	10	Some specimens purchased in fish market.
42	26 Sept.	Ecuador; Manta harbor, 4 km offshore, E of breakwater, 00°54.0'S, 80°43.0'W.	2100–2300	12	surface	night-light, dip net, poison	sand, sandy mud	sand beach	silty, green		SW 1–2	0	Sluggish flow.
43	27 Sept.	Ecuador; W of Manta, about 30 km N of La Plata Is., 00°58'S, 81°06'W, bearing 210°	2145–2430	500–600	200	5-m balloon trawl			blue		SW 1	100	
44	28 Sept.	Ecuador; La Plata Is., E side along sand beach and rock outcroppings; in tide pools and surge channels, 01°15.0'S, 81°05.0'W	1315–1500	0–1	0–1	poison, dip net	sand, rock rubble	narrow sand beach, hilly	clear, green	flooding	SW 1	100	Water temperature 22.5°C; strong surge and surf.
45	28 Sept.	Ecuador; La Plata Is., E side at anchorage about 150 m offshore, 01°16.0'S, 81°05.0'W.	2000–2200	30	surface	night-light, dip net	sand, silt, gravel	narrow sand beach, hilly	clear, green	ebbing	SW 1–2	100	Air temperature 22.0°C; water temperature 22.5°C.
46	29 Sept.	Ecuador; E of La Plata Is., 01°19'S, 81°02'W, bearing 70°	1430–1500	20–30	20–30	5-m flat trawl	clay, rubble, shell		clear, blue	flooding	SW 2	90–100	Air temperature 23–24°C; water temperature 22.8°C.
47	29 Sept.	Ecuador; between Punta Canoa and La Plata Is., 20 km E of La Plata, 01°18'S, 80°56'W, bearing 240°	1540–1630	45–50	45–50	5-m flat trawl	clay, sand, detritus		clear, blue	flooding	SW 2	100	Water temperature 22.8°C.
48	29 Sept.	Ecuador; La Plata Is., tide pools, 01°16.0'S, 81°05.0'W.	1530			purchased	tide pools	narrow sand beach, hilly					Purchased from natives.
49	29 Sept.	Ecuador; La Plata Is., E side at anchorage, about 350 m offshore, 01°16.0'S, 81°05.0'W.	2100–2330	30	surface	night-light, dip net	sand, silt, gravel	narrow sand beach, hilly	clear, blue	ebbing			Salinity 33.6‰; slight northward current.
50	29–30 Sept.	Ecuador; La Plata Is., E side at anchorage, about 150 m, offshore, 01°16.0'S, 81°05.0'W.	1945–0830	30–35	30–35	pyramidal trap	sand, silt, gravel	narrow sand beach, hilly	clear, blue	ebb low to flood high	SW 1–2	100	Net baited with dolphin meat and night-light; air temperature 22.0°C; water temperature 22.5°C.
51	30 Sept.	Ecuador; La Plata Is., E side of island, along rocks at base of cliffs, 01°16.0'S, 81°05.0'W	1500–1630	0–4	0–4	poison, dip net, spear	boulders, rubble, sand	narrow sand beach, hilly	silty, green, visibility 6 m	low	SW 1	50	Air temperature 24°C; water temperature 22.5°C; slight surge.
52	30 Sept.	Ecuador; W of La Plata Is., 01°14'S, 81°07'W at anchor on 90-m curve 5 km offshore	2000–2100	100	surface	night-light, dip net		narrow sand beach, hilly	blue, clouded by plankton	flooding	SW 1	100	
53	30 Sept.	Ecuador; W of La Plata Is., 01°14'S, 81°07'W, bearing 270°	2200–2245	200	50	5-m flat trynet, mid-water		narrow sand beach, hilly	blue, clouded by plankton	flooding	SW 1	100	
54	30 Sept.–1 Oct.	Ecuador; W of La Plata Is., 01°14'S, 81°14'W, bearing 270°	2330–0045	surface	surface	night-light, dip net		narrow sand beach, hilly	clear, blue	flooding	SW 1	100	Air temperature 20°C.
55	1 Oct.	Ecuador; 13 km N of Manta, 00°43'S, 80°41'W, bearing 270° along 40-m curve	1500–1550	20–25	20–25	5-m flat trynet		narrow sand beach, hilly	clear, blue		SW 4	0	Air temperature 25.5°C; water temperature 23.9°C; slight haze.

No.	Date	Locality	Time			Method	Bottom	Shore	Water	Tide	Wind	%	Remarks
56	1 Oct.	Ecuador; 15 km N of Manta, 00°50'S, 80°49'W.	1700–1730	75	75	5-m flat trynet	mud	—	clear, blue	—	SW 4	10–20	Air temperature 20°C; water temperature 22.4°C.
57	1–2 Oct.	Ecuador; Manta harbor, 4 km offshore, NE of breakwater 00°56.0'S, 80°42.5'W.	2345–0100	25	surface	night-light, dip net	—	sand beach	silty, green	ebbing	SW 2	100	Salinity 33.5‰; air temperature 23.5°C; water 23.0°C; abundance of *Penicillium* on bottom.
58	3 Oct.	Ecuador; La Plata Is., NE side of island 01°15.5'S, 81°05.0'W.	1130–1300	3–6	3–6	poison, dip net, spear, quinaldine	silt, coral stacks, debris	boulders, tide pools	slightly turbid	ebbing	SW 1	100	Moderate surge.
59	3 Oct.	Ecuador; La Plata Is., E side, SE of ARGOSY anchorage, 01°16.5'S, 81°04.5'W.	1345–1430	5	0–5	spear	rocks, boulders, crevices	boulders, tide pools	turbid	low	SW 1	100	Moderate surge.
60	3 Oct.	Ecuador; La Plata Is., E side, near anchorage 01°16.0'S, 81°05.0'W.	1500–1730	0–6	2–7	poison, dip net spear, quinaldine	coral stacks, silt, debris	tidepools, mountains	turbid	low	SW 1	100	Moderate surge; many alcyonarians.
61	3 Oct.	Ecuador; La Plata Is., E side, at anchorage, 01°16.0'S, 81°05.0'W.	2000–2230	30	25–30	hook and line	sandy; silt, gravel	boulders, tide pools	clear, blue	low	SW 1	—	Surface salinity 33.5‰.
62	3 Oct.	Ecuador; La Plata Is., approximately 01°16.0'S, 81°05.0'W.	—			purchase	—	—	—	—	—	—	Purchase of specimens from natives.
63	3 Oct.	Ecuador; La Plata Is., E side, at anchorage, 01°16.0'S, 81°05.0'W.	2000–2200	30	surface	night-light, dip net	sand, silt, gravel	boulders, tide pools	clear, blue	flooding	SW 1	100	Air temperature 21.5°C; water temperature 23.0°C.
64	4 Oct.	Ecuador; La Libertad, 02°13.5'S, 80°54.5'W.	—			beach seine	—	—	—	—	—	—	Purchase from natives.
65	4 Oct.	Ecuador; La Libertad, in harbor off Anglo-Ecuadorian Ltd. oil-field dock, 02°13.0'S, 80°55.0'W.	2300–2400	20	surface	night-light, dip net	sand, shells	sandy beach	turbid, green	flooding	SW 1	100	Air temperature 20.0°C; water temperature 22.5°C.
66	5 Oct.	Ecuador; 15 km E of La Plata Is., 01°15.5'S, 80°57'W.	1810–1850	35	25	5-m flat trawl, mid-water	sand, silt, gravel	boulders, tide pools	clear, blue	flooding	SW 1	—	Water temperature 22.5°C; salinity 33.4‰.
67	5–6 Oct.	Ecuador; La Plata Is., E side at anchorage, 01°16.0'S, 81°05.0'W.	2000–0115	30	surface	night-light, dip net	sand, silt, gravel	boulders, tide pools	clear, blue	flooding	SW 1	100	Water temperature 22.5°C.
68	6 Oct.	Ecuador; La Plata Is., NE part along rocks, 01°16.0'S, 81°05.0'W.	1100–1630	20	2.5–6	poison, dip net, spear	boulders, talus debris	hilly, tide pools,	visibility 6 m, blue green	ebb, low	SW 3	100	Water temperature 22.5°C; much *Padina* on rocks.
69	6 Oct.	Ecuador; La Plata Is., E side at anchorage, 01°16.0'S, 81°05.0'W.	2000–2215	30	surface	night-light, dip net	sand, silt, gravel	boulders, tide pools	clear, blue	flooding	SW 1	100	
70	6 Oct.	Ecuador; La Plata Is., approximately 01°16.0'S, 81°05'W.	"afternoon"			jigs	—	—	—	—	—	—	Purchase from local fishermen.
71	7 Oct.	Ecuador; La Plata Is., NE side, 01°15.5'S, 81°05.0'W	1000–1200	10–30	10–30	poison, dip net	coral, silt, sand in deeper water	hilly, tide pools	clear, blue	ebbing	SSW 1	10	Slight northerly current.
72	7 Oct.	Ecuador; La Plata Is., E side at anchorage, 01°16.0'S, 81°05.0'W.	1500–1520	35	25–35	spear, hand	silt, sand	tide pools, hilly	rather clear, blue	low	SW 1	50	Strong northerly current below thermocline (at 35m), no current above; surface temperature 23.5°C; at 30 m 19.0°C; at 33 m.
73	8 Oct.	Ecuador; La Plata Is., E side, along sandy beach, to 30 m offshore, 01°16.0'S, 81°05.0'W.	1100–1200	35	0–1.5	20-m bag seine, poison	fine sand, rubble, gravel	sand beach	clear, white	flooding after low	SW 1	50	Water temperature 23.3°C; air temperature 30.0°C; salinity 33.5‰; slight southerly current.
74	8 Oct.	Ecuador; La Plata Is., E side, along tide pool, N of sandy beach (Collection Sta. 73), 01°16.0'S, 81°05.0'W.	1200–1215	0–<.5	0–<.5	poison	debris	tide pools, hilly	clear	flooding	SW 1	50	Salinity 34.9‰; air temperature 30.0°C.
75	8 Oct.	Ecuador; mainland coast at Salango Island, in surf, 01°35.5'S, 80°53.5'W.	1100–1630	—	surface	hook and line	—	rocky	slightly silty, blue	flood high	SW 2	—	Caught by Alfred C. Glassell, Jr.
76	8 Oct.	Ecuador; La Plata Is., E side SE of anchorage 180 m offshore, 01°16.5'S, 81°04.0'W.	1530–1700	10	8–10	poison, dip net	silt over coral heads, rubble, detritus	tide pools, hilly	slightly silty, blue	flood high	SW 1–2	90	Air temperature 28.0°C; water temperature 23.0°C.
77	8 Oct.	Ecuador; La Plata Is., NE side, N of anchorage, along rocks and offshore, 01°16.0'S, 81°05.0'W	2000–2230	5–35	surface	night-light, dip net	sand, coral	tide pools, hilly	clear, blue	ebbing	SW 1	0–100	
78	9 Oct.	Ecuador; La Plata Is., NE side, about 15 m offshore in reefs, 01°15.5'S, 81°05.0'W.	1100–1330	10	5–6	spear, poison	silt, sand, coral heads	tide pools, hilly	slightly turbid, blue	ebb low	SW 2	10	Bright sky.

TABLE 1—Continued

Sta. No.	Date	Location	Time, EST	Depth of water, meters	Depth of capture, meters	Method of capture	Bottom	Shore	Water	Tide	Wind dir. & force, Beaufort	Cloud Cover, %	Remarks
79	9 Oct.	Ecuador; La Plata Is., E side, at anchorage, 01°16.0'S, 81°05.0'W.	2000–2200	35	surface	night-light, dip net	sand	tide pools, hilly	clear, blue	flooding	W-N 1	100	Moderate southerly current.
80	9–10 Oct.	Ecuador; La Plata Is., E side at anchorage, 01°16.0'S, 81°05.0'W.	2000–1000	30	30	pyramidal fish trap	sand, silt	tide pools, hilly	clear, blue		SW 1	100	—
81	9–10 Oct.	Ecuador; La Plata Is., E side, rocky tide pool area, 01°16.0'S, 81°05.0'W.	2000–1000	5	0–5	multimeshed gill net	sand, coral	tide pools, hilly	clear, blue		SW 1	100	Net stretched between coral heads.
82	9–10 Oct.	Ecuador; La Plata Is., E side at anchorage, 01°16.0'S, 81°05.0'W.	2000–0800	<20	<20	25-hook trotline	sand	tide pools, hilly	clear, blue		SW 1	100	Most of line lost.
83	10 Oct.	Ecuador; La Plata Is., E side at anchorage, 01°16.0'6, 81°05.0'W.	1300–1530	10–25	10–25	spear	silt, sand, coral	tide pools, rocks	clear, blue	ebbing	SW 2	100	3-knot northward current; visual thermocline at 25 m.
84	10 Oct.	Ecuador; La Plata Is., approximately 01°16.0'S, 81°05'W.				beach seine							Purchased from native fishermen.
85	10 Oct.	Ecuador; La Plata Is., E side at anchorage, 01°16.0'S, 81°05.0'W.	2000–2200	35	surface	night-light, dip net	sand	tide pools, hilly	clear, blue		SW 1	100	Water temperature 23.5°C.
86	10–11 Oct.	Ecuador; La Plata Is., E side, N of anchorage in rocks, 01°16.0'S, 81°05.0'W.	1000 on 10/10 to 1400 on 10/11	5–6	5–6	pyramidal fish trap	sand	tide pools, hilly	clear, blue				Trap wedged in coral reef.
87	11 Oct.	Ecuador; La Plata Is., E side along sandy beach and in tide pools, 01°16.0'S, 81°05.0'W.	1300–1415	0–0.2	0–0.2	hand; BB gun	tide pool	rocky tide pool, beach	clear	low	SW 3	100	Considerable surf.
88	12 Oct.	Ecuador; 15 km W of La Plata Is., off continental shelf, 01°15'S, 81°14'W.	0415–0600	125	surface	night-light, squid jigs, dip net			clear, blue		SW 2	100	Water temperature 22.4°C; sea SW 2.
89	12 Oct.	Ecuador; N of Salinas, SW of Salingo, approximately, 01°45'S, 80°59'W.	1200–1400		surface	hook and line			slightly turbid, blue				Caught by Alfred C. Glassell, Jr., aboard *Sea Quest*.
90	14 Oct.	Ecuador; La Libertad, sea buoy (drying on dock) 02°18.5'S, 80°55.0'W.				scraped by hand							Specimens kept dry.
91	11 Oct.	Panama; Gulf of Panama, Rio Chinina.				shrimp boat trawl							Material from Edward Klima of the Inter-American Tropical Tuna Commission, Panama.

TABLE 1 (Continued) *Plankton Collections*

Sta. No.	Date	Location	Time, EST	Depth of water, meters	Depth of capture, meters	Method of capture	Bottom	Water	Water Temp.	Tide	Wind dir. & force, Beaufort	Cloud Cover, %	Remarks
P-1	7 Sept.	Panama; Panama Bay, WNW of Las Perlas Islands, 08°48'N, 79°25'W.	0730-1130	—	surface	high-speed plankton sampler	—	greenish blue, slightly turbid	—	—	SW 0-1	80-100	Vessel speed 8 knots[1]; sampler fishing too shallow.
P-2	8 Sept.	Panama; Panama Bay, from 07°57'N, 78°45'W to 07°50'N, 78°50'W.	1000-1100	—	surface	high-speed plankton sampler	—	greenish blue, slightly turbid	27.5	—	SW 0-1	0	Sampler fishing too shallow.
P-3	8 Sept.	Panama; Panama Bay, 07°41'N, 78°20'W.	1130-1330	—	surface	high-speed plankton sampler	—	blue green	27.5	—	SW 0-1	0-80	—
P-4	13 Sept.	Panama; Panama Bay, <6 km WSW of Bahia Piñas to WSW of Bahia Piñas, bearing 270°. 07°32'S, 78°16'W to 07°2'8S, 78°22'W.	0850-0950	—	1.5-3	high-speed plankton sampler	—	blue green	—	—	—	—	—
P-5	13 Sept.	Panama; Panama Bay, <90 km WSW of Bahia Piñas, to <45 km WSW of Bahia Piñas, 07°26'S, 78°26'W to 07°22'S, 78°33'W.	1000-1100	—	1.5-3	high-speed plankton sampler	—	blue water	27.5	—	—	—	Sampler lost on log after this sample taken.
P-6	13 Sept.	Panama; approximately 75 km WSW of Bahia Piñas, 07°17'N, 78°42'W.	1115-1215	—	—	high-speed plankton sampler	—	blue water	—	—	—	—	—
P-7	16-17 Sept.	Colombia; W. of Bahia Solano, bearing 180°, about 15 km W of Punta San Francisco Solano, 06°18.5'N, 77°34.0'W.	2445-0115	<750	surface	1-meter plankton net, 00 nylon	—	rather turbid	—	flood, nearly high	SW 1	0	1-knot northerly current; net towed at surface with ca 1/3 of ring emerging; vessel speed 3 knots.[2]
P-8	17 Sept.	Colombia; between Punta Uría and Buenaventura, approximately 03°57'N, 77°35'W.	2000-2030	>35-75	surface	1-meter plankton net, 00 nylon	—	—	—	—	—	—	—
P-9	23 Sept.	Colombia; <10 km S of Gorgona Is., 02°39'N, 78°38'W.	2000-2030	<100	surface	1-meter plankton net, 00 nylon	—	clear, with detritus	27.0	flood	SW 1	100	—
P-10	26 Sept.	Ecuador; NNE of Manta, off Cabo Pasado, 00°12'S, 80°36'W.	1140-1200	75	surface	1-meter plankton net, 00 nylon	—	clear, blue	23.5	—	SW 2	100	Salinity 33.3‰; air temperature 22.6°C.
P-11	27 Sept.	Ecuador; W of Manta, about 30 km N of La Plata Is., 00°58'S, 81°06'W, bearing 210°	2300-2345	—	surface	1-meter plankton net, 00 nylon	—	—	—	—	—	100	—
P-12	30 Sept.	Ecuador; W of La Plata Is., at >180-m curve, 01°14'S, 81°14'W, bearing 270°	2204-2234	>180	surface	1-meter plankton net, 00 nylon	—	blue; slightly turbid with plankton	22.5	flooding	SW 1	100	Sea SW 1.
P-13	1 Oct.	Ecuador; 13 km N of Manta, along 35-m curve, 00°43'S, 80°41'W, bearing 270°	1445-1515	35	surface	1-meter plankton net, 00 nylon	—	clear, blue	23.9	—	—	0	Air temperature 25.5°C; tow very poor, few organisms.
P-14	1 Oct.	Ecuador; 15 km N of Manta, 00°50'S, 80°49'W.	1645-1715	75	surface	1-meter plankton net, 00 nylon	—	clear, blue	—	—	SW 4	10-20	Tow combined with P-13. SW swell 2-3.
P-15	5 Oct.	Ecuador; 15 km E of La Plata Is., 01°15'S, 80°57'W.	1810-1850	35	surface	1-meter plankton net, 00 nylon	—	clear, blue	22.5	—	SW 1	—	Salinity 33.4‰.

[1] Vessel speed 8 knots for samples P-1 through P-6.
[2] Vessel speed 3 knots for samples P-7 through P-15.

TABLE 2. *Weather Log of Argosy, September 5 to October 13, 1961**

Time, EST	0000	0400	0800	1200	1600	2000
Sept. 5	Cl, Cle, cool, S–4	Cloudy	Cl	Cl	Cl	R, S–4
Sept. 6	Calm	Cl, S–2	DR, SSE–3	R, SQ, SE–4	S–1	Calm
Sept. 7		Cl, SE–1	Cl, ESE–2	Cl, S–4	OV, SE–3	E–1
Sept. 8	OV, Calm	Cl, E–1	Cl, SE–2	Cl, SSE–4	Cl, S–2	Calm, SSW–1
Sept. 9	Cl, Calm	Cl, SW–1	Fair & Cle, W–2	Cl, NW–3	R, SE–1	OV, R, SE 1–2
Sept. 10	R, Calm	R, SE–2	OV, R, W–2	R, SSE–2	Sun, SSE–1	OV, R, NNW–1
Sept. 11	R, Calm	Cle, SE 0–1	Cle, E–1	Cl, SSE–3	Cl, S–3	DR, SSE–2
Sept. 12	DR, SE–1	DR, W–1	DR, SW–1	Cl, SSW–3	Cl, SW–2	Cl, NE–1
Sept. 13	Cle, E–1	Cle, SE–1	Cl, Calm	Cle, S–2	Cle, SW–1	Cle, SW–1
Sept. 14	Cle, SW–1	OV, DR, Calm	R, S–1	R, SW–4	Cl, Cle, SW–3	Cl, Cle, W–1
Sept. 15	Cl, Cle, W–1	R, SE–2	DR, SE 1–2	DR, Calm	OV, DR, Calm	OV, SE–1
Sept. 16	Cl, Cle, S–2	OV, DR, S–1	Cl, Cle, SW–2	Cl, Cle, S–1	Cl, Cle, W–1	Cl, Cle, SSW–2
Sept. 17	Cl, Cle, S–3	R, E–1	Cl, S–1	Cl, SW–2	Cl, SSW–2	Cl, SSW–1
Sept. 18	Cl, Calm	Foggy, Calm	OV, Calm	Cle, S–1	R, NW–2	DR, Calm
Sept. 19	Cl, Calm	Dr, Calm	Cl, SE–1	OV, S–3	OV, DR, SW–4	OV, DR, SSW–3
Sept. 20	R, Calm	R, S–1	OV, DR, S–3	Cl, S–3	Cl, NW–1	Cl, WNW–1
Sept. 21	R, SW–1	R, S–1	R, SE–2	OV, S–1	Cl, Cle, SSW–3	Cl, Cle, SW–3
Sept. 22	Cl, Cle, SW–3	Cl, SSW–1	R, E–1	R, E–1	OV, NW–1	OV, R, S–1
Sept. 23	OV, S–1	DR, S–1	Cl, Cle, SW–1	Cl, Cle, W–2	OV, W–3	OV, SW–3

Sept. 24	OV, SW-3	OV, SW-2	Cl, SW-1	Cle, WSW-3	Cl, SW-2
Sept. 25	Cl, SW-2	OV, SW-2	OV, SW-1	OV, SW-2	Cle, S-1
Sept. 26	Calm, Cle	Cl, Cle, S-2	OV, S-1	OV, W-3	Cl, Cle, SW-4
Sept. 27	OV, SSW-1	OV, S-2	OV, S-1	Cle, W-3	OV, S-1
Sept. 28	OV, S-1	OV, S-2	OV, S-2	Cl, Cle	OV, S-1
Sept. 29	OV, S-1	OV, S-1	OV, S-1	OV, S-2	OV, S-1
Sept. 30	OV, SSW-1	OV, SW-1	OV, SW-2	Cl, Cle, SW-2	Cl, Cle, SW-2
Oct. 1	Cl, Cle, SW-2	OV, SW-1	OV, SW-1	Cle, SSW-3	Cl, SW-2
Oct. 2	Cle, SSW-1	OV, S-1	OV, SW-1	OV, W-1	OV, Calm
Oct. 3	OV, Calm	OV, S-1	OV, SW-2	OV, SW-2	OV, SW-1
Oct. 4	OV, S-1	OV, SW-1	OV, SW-2	OV, SW-3	OV, SW-1
Oct. 5	OV, NNW-1	OV, Calm	OV, WSW-2	OV, SW-1	OV, SW-1
Oct. 6	OV, Calm	Cle, OV, SSW-2	Cl, SSE-1	Cl, SW-3	OV, SW-1
Oct. 7	OV, SW-1	OV, SW-1	OV, SW-1	Cl, SW-1	OV, SW-1
Oct. 8	OV, Calm	OV, SW-1	Cl, Cle, S-1	Cle, S-1	Cl, Cle, S-1
Oct. 9	OV, SW-2	OV, SW-1	OV, SW-2	OV, SW-2	OV, SW-1
Oct. 10	DR, Calm	DR, SW-1	OV, SSW-1	Cl, Cle, SW-2	OV, SW-1
Oct. 11	OV, WSW-1	OV, SW-3	OV, SSW-4	Cl, SSW-4	OV, SW-1
Oct. 12	OV, WSW-1	OV, SW-2	OV, SSW-1	OV, SW-2	OV, SW-1
Oct. 13	OV, SW-2	OV, SW-1	OV, SW-2	Cl, Cle, WSW-3	OV, WSW-1

* Abbreviations: Cl = Cloudy; Cle = Clear; DR = Drizzle; R = Rain; SQ = Squalls; OV = Overcast

TABLE 3. *Data for Argosy oceanographic stations occupied in the coastal waters of Ecuador, October, 1961*

Station No.	H-1	H-2	H-4	H-5	H-6	H-7	H-8	H-9	H-10	H-12	H-13	H-14	H-15
Date	2 X 61	2 X 61	2 X 61	2 X 61	2 X 61	5 X 61	5 X 61	5 X 61	5 X 61	12 X 61	12 X 61	12 X 61	12 X 61
Latitude	01°04'20"S	01°06'40"S	01°14'00"S	01°13'50"S	01°13'30"S	01°50'S	01°41'S	01°32'S	01°21'S	01°14'S	01°32'S	01°47'05"S	02°14'30"S
Longitude	80°56'30"W	80°57'50"W	81°02'30"W	81°06'40"W	81°10'20"W	81°15'W	81°08'W	81°07'W	80°55'W	80°10'W	81°07'W	80°56'05"W	80°57'W
Time, GMT	1910	1945	2100	2130	2200	1015	1245	1435	1600	0535	0840	1200	1400
Depth to bottom, m	80	95	100	120	400	900	200	—	—	—	—	—	—
Surface temp., °C	23.5	23.4	23.0	23.0	22.9	22.5	22.9	23.0	23.0	22.4	22.8	23.0	23.2
Weather	Fair, sl. haze	Overcast	Fair & clear	Fair	Fair	Overcast, bright	Overcast, bright	Overcast, bright	Overcast	Overcast	Overcast	Overcast, bright	Overcast
Wind & force	SW>2	SW>1	SW 1	SW 1	SW 1	SW 1	SW 1–2	SW 2	SW 1	SW>2	SW 2	SW>1	SW 1
Sea & swell	SW<1, sl.	SW<1	SW 1	SW 1	SW 1	SW 1	SW sl.	SW 1	SW 1 sl.	SW 2	SW 2	SW>1	SW 1
Type of clouds, %	S Cu[1] 100	S Cu 100	S Cu 100	S Cu 100	S Cu 100	S Cu 100	S Cu 100	S Cu 100	S Cu 100	S Cu 100	S Cu 100	S Cu 100	S Cu 100
Air temp., °C	21.5	21.0	—	—	20.2	24.0	27.0	24.0	—	21.5	21.5	25.0	—
Barometer, mm	29.88	—	—	—	—	29.86	29.83	29.78	29.80	29.87	29.82	—	29.82
Salinity, ‰													
Depth, m 0	—	33.47	33.61	33.58	33.57	33.61	33.63	33.60	33.46	33.52	33.57	33.57	33.86
7	33.55	—	—	—	—	—	—	33.50	33.54	—	—	33.54	—
15	—	—	—	—	34.97	33.65	33.59	33.58	33.56	33.47	33.88	33.70	—
21	—	—	—	—	—	—	—	34.02	33.78	—	—	34.34	—
30	—	—	—	—	—	34.09	34.05	33.61	33.54	34.37	—	34.89	—
37	—	—	—	—	—	—	—	—	—	—	—	35.04	—
45	—	—	—	—	34.59	34.81	35.01	—	—	35.06	34.93	34.97	—
52	—	—	—	—	—	—	—	—	—	—	—	34.98	—
60	—	—	—	—	34.95	34.99	35.02	—	—	35.02	34.99	—	—
75	35.09	—	—	—	—	—	—	35.00	—	—	34.98	—	—
137	—	—	—	—	—	34.99	34.99	—	—	—	—	—	—
152	—	—	—	—	—	—	—	—	—	—	—	—	—

[1] Stratocumulus

Bibliography

The present report had been largely completed when, in December 1967, a fire destroyed the office of the writer. While the original logbook and station data were recovered essentially intact, an extensive, annotated bibliography dealing with major references on the Panama Bight was destroyed, and only citations to some of the original references were recovered. These references cover a wide variety of descriptions and documentation of the natural history of northwestern South America, and include a number of rare yet germane references to the area. They represent many hours work in the Library of Congress, the library of the Academy of Natural Sciences of Philadelphia, and the library of the American Museum of Natural History, and are included because some references will hopefully be of assistance to contributors to *Argosy* reports, and also to others interested in the areas. Because of the writer's interest in fishes, this section is more heavily stressed than other areas, but attempts were made to include references which would be meaningful to a variety of workers. In the interest of time, and because it is impractical to try to correct and reannotate each reference and to complete missing pagination or otherwise partially complete references, the author felt it expedient to present these incomplete, unannotated references for what they may be worth. For convenience of use, the bibliography has also been listed below by broad categories.

SELECTED REFERENCES ON THE NATURAL HISTORY OF NORTHWESTERN SOUTH AMERICA INCLUDING THE PANAMA BIGHT

General Descriptions or Narratives: Bates, 1947; Blanchard, 1962; Bockmann, 1941; Bürger, 1919; Cepolla, 1929; Enock, 1914; Ferdon, 1940–1941; Heatherington, 1846; James, 1942; Larrea, 1924 *et seq.*; Lasso, 1944; Lévine, 1914a, b; Linke, 1954, 1960; Maull, 1937; Merizalde del Carmen, 1921; Müller, 1935; Grace Murphy, 1943; R. G. Murphy, 1925a *et seq.*; Onffroy de Thoron, 1866; Pariseau, 1963; Platt, 1956, 1959; Preuss, 1914; Reyes, 1956; Robinson, 1895; Rogers, 1712; Scruggs, 1901; Troll, 1931; Ulloa, 1771; U.S. Naval Oceanographic Office, 1965; Vanderbilt, 1927; de Velasco, 1841, 1958; von Hagen,

1940 *et seq.*; Voss, 1967; Wafer, 1699, 1706; Wagner, 1861; Walker, 1822a, b; Whitaker, 1948.

Historical Travels: Alborñoz, 1854; Caldes y Tenorio, 1936; Cieza de León, 1864; Colnett, 1798; la Condamine, 1778; Dampier, 1717; Darwin, 1832–1836 *et seq.*; Humboldt, 1805 *et seq.*; Larrea, 1948–1952 *et seq.*; Onffroy de Thoron, 1866; Reyes, 1956; Rogers, 1712; Saville, 1917; Sievers, 1914, 1931; Ulloa, 1771; de Velasco, 1841, 1958; von Hagen, 1940 *et seq.*; Wafer, 1699, 1706.

Geography: Bengtson, 1924; Bennett, 1925; Blanchard, 1962; Botero, 1950; Bruno, 1935; Cornish, 1955; Enock, 1914; Ferdon, 1940–1941, 1950; Fermín Ceballos, 1888; Gierloff-Emden, 1959; Goez, 1947; James, 1942; Knoche, 1932; Larrea, 1948–1952 *et seq.*; Lasso, 1944; Lévine, 1914a, b; Linke, 1954, 1960; Lleras Codazzi, 1926; López, 1907; Maull, 1937; Mendozo Nieto, 1942; Merizalde del Carmen, 1921; Mosquera, 1953; Müller, 1935; Grace Murphy, 1943; R. C. Murphy, 1925a *et seq.*; Pariseau, 1963; Paz y Miño, 1950; Platt, 1956, 1959; Robinson, 1895; Sachet, 1962a, b; Scruggs, 1901; Sheppard, 1930b; Sievers, 1931; Torres Mariño, 1938; Troll, 1931; Ulloa, 1771; U. S. Naval Oceanographic Office, 1965; Vanderbilt, 1927; von Hagen, 1940 *et seq.*; Voss, 1967; Wafer, 1699, 1706; Wagner, 1861; Walker, 1822a, b; Whitaker, 1948; Wolf, 1869–1876 *et seq.*

Government and Politics: Enock, 1914; Heatherington, 1846; James, 1942; Lasso, 1964; Lévine, 1914a, b; Linke, 1959, 1960; Onffroy de Thoron, 1866; Pariseau, 1963; Reyes, 1956; Scruggs, 1901; de Velasco, 1841, 1958; von Hagen, 1940 *et seq.*; Walker, 1822a, b; Whitaker, 1948.

Anthropology and Archeology: Archer, 1937; Bennett, 1944; Bollaert, 1860; Collier and Murra, 1943; Coon, 1954; Cuervo Marquez, 1917; Dorsey, 1901; Führmann and Mahor, 1914; Gilmore, 1950; Larrea, 1948, 1952 *et seq.*; Maull, 1937; Merizalde del Carmen, 1921; R. C. Murphy, 1925a *et seq.*; Ortíz, 1937; Pariseau, 1963; Perez de Barradas, 1954; Preuss, 1914, 1929; Recasens and Oppenheim, 1943–1944; Saville, 1903 *et seq.*; Schottelius, 1941; von Hagen, 1940 *et seq.*; Voss, 1967; de Wavrin, 1936; West, 1951; Whitaker, 1948.

Climate: Anonymous, 1938, 1956a; Annuario Meteorológico, 1955; Chapel, 1927; Dickerson, 1917; Franze, 1927; Henry, 1922; Heatherington, 1846; Jennings, 1944; Knoch, 1930; Knoche, 1932; Larrea, 1948–1952 *et seq.*; Lasso, 1944; Lévine, 1914a, b; Linke, 1954, 1960; Müller, 1935; R. C. Murphy, 1925a *et seq.*; Pariseau, 1963; Portig, 1965; Sheppard, 1930b, 1933; Torres Mariño, 1938; Troll, 1931; U. S. Naval Oceanographic Office, 1965; Voss, 1967; Wagner, 1861; Walker, 1822a, b.

Geology and Palaeontology: F. M. Anderson, 1927, 1928; J. L. Anderson, 1945; Baur, 1891, 1897; Berry, 1929; Bosworth, 1922; Chubb, 1925; Crossland, 1927; Darwin, 1844; Davies, 1929; Dickerson, 1917; Durham and Allison, 1960; Ekman, 1953; Gansser, 1950; Granja, 1957; Gregory, 1930; Heacock and Worzel, 1955; Hill, 1898; Iddings and Olssen, 1928; Jones, 1950; Karsten, 1858, 1886; Larrea, 1924 *et seq.*; Li, 1930; Lleras Codazzi, 1926; Maack, 1874; Mac-Donald, 1919; R. C. Murphy, 1925a *et seq.*; Neale, 1866; Nygren, 1950; Olssen, 1931 *et seq.*; Oppenheim, 1949, 1952; Rubio y Muñoz-Bocanegra, 1949; de la Rüe, 1933, 1934; Sauer, 1950; Sheppard, 1927a *et seq.*; Sinclair and Berkey,

1923; Stübel, 1906; Terry, 1941, 1956; Vaughan, 1919; White, 1927; Wilson, 1886; Wolf, 1869–1876 *et seq.*

Natural Resources: Anonymous, 1937; Dawson, 1963; Enock, 1914; Ferdon, 1940–1941, 1950; Henn, 1914; Larrea, 1948–1952 *et seq.*; Lasso, 1944; Lévine, 1914a, b; Maack, 1874; Maull, 1937; R. C. Murphy, 1925a *et seq.*; Onffroy de Thoron, 1866; Ortiz Borda, 1961; Pariseau, 1963; Pérez Arbeláez, 1953; Posada y Aranga, 1909; Rioja, 1962; Sachet, 1962a, b; UNESCO, 1961; Uribo, 1935, 1936; von Hagen, 1940 *et seq.*; Voss, 1967; Williamson, 1918.

General Scientific Expeditions and Journeys: Agassiz, 1892, 1906; Beebe, 1938; Caruccio, 1886; Chubb, 1925; Coventry, 1944; Dawson and Beaudette, 1959; de Sylva, 1963; Festa, 1901; Fraser, 1943; Gigliogli, 1875; Larrea, 1948–1952 *et seq.*; Meredith, 1939; Grace Murphy, 1943; R. C. Murphy, 1925a *et seq.*; Murray, 1895; Sievers, 1914; Targinoni-Tozzetti, 1877; Vanderbilt, 1927; von Hagen, 1940 *et seq.*; Voss, 1967.

Oceanography: A. W. Anderson, 1963; Anonymous, 1938, 1939, 1948, 1956b, 1962; Austin, 1960; Bennett, 1963, 1966; Blackburn, 1966; Blackburn *et al.*, 1962; Brandhorst, 1958; Broenkow, 1965; Cromwell, 1958; Cromwell and Bennett, 1959; Fisher, 1958; Fleming, 1938, 1940; Forsbergh, 1963; Forsbergh and Broenkow, 1965; Forsbergh and Joseph, 1964; Gierloff-Emden, 1959; Gilmartin, 1964; Gunther, 1936; Holmes *et al.*, 1958; Hubbs and Roden, 1964; Hubbs and Rosenblatt, 1961; Kirkpatrick, 1926; Larrea, 1948–1952 *et seq.*; Marmer, 1930; Menard, 1960; R. C. Murphy, 1925a *et seq.*; Ortiz Borda, 1961; Roden, 1962, 1963; Shaefer *et al.*, 1958; Schmidt, 1925; Schott, 1932; Schweigger, 1958; Smayda, 1963; Smith *et al.*, 1964; Sund, 1959; U. S. Naval Oceanographic Office, 1965; Vegas, 1963; Voss, 1967; Wooster, 1959; Wooster and Cromwell, 1958; Wyrtki, 1964 *et seq.*

Zoology (*general*): Bangs *et al.*, 1905; Berry, 1959; Ekman, 1953; Faxon, 1893; Gigliogli, 1875; Gilmore, 1950; Henn, 1914; Klawe, 1964; Knoche, 1932; Larrea, 1948–1952 *et seq.*; Maack, 1874; R. C. Murphy, 1925a *et seq.*; Orces, 1942; Posada y Aranga, 1909; Rioja, 1962; Thayer and Bangs, 1905; Uribe, 1935, 1936; Voss, 1967; Williamson, 1927.

Fishes: Abbott, 1899; Angelescu, 1960, 1961; Behre, 1928; Böhlke and Robins, 1968; Borodin, 1928, 1932; Boulenger, 1887 *et seq.*; Breder, 1925 *et seq.*; Briggs, 1961; de Buen, 1952, 1960; Clemens, 1955; Clemens and Nowell, 1963; Corlett, 1957; de Sylva, 1961, 1963; Díaz, 1965; Eigenmann, 1905 *et seq.*; Eigenmann and Allen, 1942; Eigenmann and Henn, 1914; Eigenmann *et al.*, 1914; Evermann and Goldsborough, 1909; Farrington, 1942, 1953; Fowler, 1938 *et seq.*; Garman, 1877 *et seq.*; Gilbert, 1890, 1891; Gilbert and Starks, 1904; Gill, 1864; Günther, 1861, 1869; Halstead, 1953; Henn, 1914; Herre, 1936; Hildebrand, 1938 *et seq.*; Hoedeman, 1960; Howard and Ueyanagi, 1965; Hubbs, 1953, 1959; Hubbs and Roden, 1964; Hubbs and Rosenblatt, 1961; Hunter and Mitchell, 1966; Jordan 1886 *et seq.*; Kendall and Radcliffe, 1912; Kner and Steindachner, 1865; Koepcke, 1951, 1959; Larrea, 1948–1952 *et seq.*; Meek and Hildebrand, 1923–1928; Miles, 1942; R. C. Murphy, 1925a *et seq.*; Myers, 1941; Myers and Wade, 1946; Nichols and Murphy, 1944; Orces, 1947 *et seq.*; Ortiz Borda, 1961; Osburn, 1916; Regan, 1906–1908; Rendahl, 1941; Ricker, 1959; Rosenblatt, 1963; Rosenblatt and Baldwin, 1958; Rosenblatt and

Walker, 1963; Rubinoff, 1963; Schaefer, 1955; Schmitt and Schultz, 1940; Seale, 1940; Snodgrass and Heller, 1903 *et seq.*; Starks, 1906; Steindachner, 1875a *et seq.*; Vanderbilt, 1927; Voss, 1967; Wagner, 1865a, b; Walford, 1937; Wilson, 1916.

Invertebrates: Adams, 1952; Berry, 1959; Crossland, 1927; Dana, 1852; Davies, 1927; Faxon, 1893; Finnegan, 1931; Garth, 1948; Hertlein and Strong, 1955; Hubbs and Roden, 1964; Hubbs and Rosenblatt, 1961; Larrea, 1948–1952 *et seq.*; Loesch and Avila, 1964; Milne-Edwards, 1868–1897, 1875–1880; Morch, 1859–1861; R. C. Murphy, 1925a *et seq.*; Nobili, 1901; Pesta, 1931; Sivertsen 1933; Voss, 1967.

Birds: Chapman, 1926; Larrea, 1948–1952 *et seq.*; R. C. Murphy, 1925a *et seq.*; Robins, 1958; Thayer, 1905; Thayer and Bangs, 1905.

Marine Mammals: Clarke, 1957 *et seq.*; Larrea, 1948–1952 *et seq.*; R. C. Murphy, 1925a *et seq.*

Botany: Dawson, 1963; Eggers, 1892; Holdridge *et al.*, 1960; Larrea, 1948–1952 *et seq.*; R. C. Murphy, 1925a *et seq.*; Svenson, 1936.

Commercial Fisheries: Anonymous, 1961, 1963, 1967; Boerema, 1960; Butler, 1965; Doran, 1953; Eisenmann, 1958; Ellis, 1961; Fischer, 1966; Food and Agricultural Organization, 1960; Greenback, 1960; Hagberg *et al.*, 1967; Howard and Godfrey, 1951; Institute of Marine Resources, 1965; Jansen, 1944; Larrea, 1948–1952 *et seq.*; Lindner, 1957; Miles, 1942; R. C. Murphy, 1925a *et seq.*; Osorio-Tafall, 1951; Oswald, 1964; Paz Andrade, 1956; Quiroga, 1964; Quiroga and Orbes, 1963a *et seq.*; Quiroga-Ríos, 1958; Ripley, 1964; Ruiz, 1966; Saenz, 1962; Schaefer, 1955; Voss, 1967; Wilton, 1949.

ABBOT, J. F.
   1899.   The marine fishes of Peru. Proc. Acad. nat. Sci. Philad., (1899):324–364.

ADAMS, C. B.
   1852.   Catalogue of shells collected at Panama with notes on synonymy, station, and habitat. Ann. Lyc. Nat. Hist., N.Y., *5*:229–344.

Agassiz, Alexander
   1892.   General sketch of the expedition of the ALBATROSS from February to May, 1891. Bull. Mus. comp. Zool. Harv., *23*:1–85, 22 pls.
   1906.   Report on the scientific results of the expedition to the eastern tropical Pacific . . . ALBATROSS. 5. General report of the expedition. Mem. Mus. comp. Zool. Harv., *33*:1–75, 96 pls.

ALBORÑOZ, G. HUGO L.
   1854.   Por tierras ecuatorianas.

ANDERSON, A. W.
   1963.   Oceanographic survey in Central and South American waters. Publ. U.S. hydrogr. Off., (05562).

ANDERSON, FRANK M.
   1927.   The marine Miocene deposits of north Colombia near Puerto Colombia at Tubera Mountain. Proc. Calif. Acad. Sci., 4th Ser., *16*(3):87–95, pls. 2,3.

1928. Notes on lower Teritiary deposits of Colombia and their molluscan and foraminiferal fauna. Proc. Calif. Acad. Sci., *17*(1):1–29, 11 figs.

ANDERSON, J. L.
1945. Petroleum geology of Colombia. Bull. Am. Ass. Petrol. Geol., *29:*1065–1142.

ANGELESCU, V.
1960. Estado actual de los conocimientos sobre las migraciones de peces marinos en los paises latinoamericanos. II. Sector Pacífico: Centroamérica. UNESCO–Symposium sobre migraciones de organismos marinos en Montevideo, pp. 12–15.
1961. Estado actual de las investigaciones en biología de peces en América Latina. Sem. Latinoamer. Est. Ocean., Concepción, Chile, *14:*13–17.

ANONYMOUS
1937. Informe científico sobre la región Quibidó–Buenaventura. Minería, *5:*4550–4565.
1938. Handbuch der Westküste Amerikas. II. Teil: Peru, Ecuador, Colombia, Panama, Costa Rica, Nicaragua, Honduras, El Salvador, Guatemala. Dt. hydrogr. Inst., Hamburg.
1939. Station lists for VELERO cruise relating to work of Joseph A. Mcculloch. Allan Hancock Pacif. Exped., *6:*3–30, 146–152, 180–184, 232–234.
1948. Surface water temperatures at Coast and Geodetic Survey tide stations, Pacific Ocean. Publ. U.S. Cst. geod. Surv., TW *2:*42 pp.
1956a. Monthly meteorological charts of the eastern Pacific Ocean. Publ. British meteorol. Off., *518:*122 pp.
1956b. Surface water temperatures at tide stations: Pacific coast of North and South America and Pacific Ocean islands. Publ. U.S. Cst. geod. Surv., 280, 5th ed.
1961. Ecuador turning to sea resources. Andean Air Mail and Peruvian Times, *21*(1050): 8.
1962. Surface water temperature and salinity, Pacific coast, North and South America and Pacific Ocean islands. Publ. U.S. Cst. geod. Surv., 31–33, 1st ed.
1963. Fisheries of Ecuador. Market News Leafl., U.S., (82), 12 pp.
1967. Ecuador—fishing industry grows. Comm. Fish. Rev., *29*(3):20.

ANUARIO METEOROLÓGICO
1955. Reports for 1949; 1950–1951; 1952–1954. Ministério de Agricultura, Bogotá.

ARCHER, W. A.
1937. Exploration in the Chocó Intendency of Colombia. Sci. Monthly, *44:*418–434; 457.

AUSTIN, THOMAS S.
1960. Oceanography of the east central equatorial Pacific as observed during Expedition Eastropic. Fish. Bull., U.S., *60*(168):257–282.

BANGS, OUTRAM, THOMAS BARBOUR, WILMOT W. BROWN, JR., AND JOHN E. THAYER
  1905. The vertebrata of Gorgona Island, Colombia. Bull. Mus. comp. Zool. Harv., *46*(5):87–102.

BATES, NANCY BELL
  1947. East of the Andes and west of nowhere. Charles Scribner's Sons, N.Y., 237 pp.

BAUR, G.
  1891. On the origin of the Galápagos Islands. Am. Nat., *25:*217–229; 307–326.
  1897. Observations on the origin of the Galápagos Islands, with remarks on the geological age of the Pacific Ocean. Am. Nat., *31:*262.

BEEBE, WILLIAM
  1938. Eastern Pacific expeditions of the New York Zoological Society 14. Introduction, itinerary, list of stations, nets, and dredges of the eastern Pacific ZACA expedition, 1937–38. Zoologica, N.Y., *23*(3): 287–298, 2 text figs.

BEHRE, E. H.
  1928. A list of the fresh water fishes of western Panama between 8°41′N and 83°15′W. Ann. Carneg. Mus., *18:*305–328.

BENGTSON, N. A.
  1924. Some essential features of the geography of the Santa Elena Peninsula, Ecuador. Ann. Ass. Am. Geogr., *14*.

BENNETT, E. B.
  1963. Un atlas oceanográfico del Océano Pacífico oriental tropical, basado en los datos de la expedición EASTROPIC, Oct.–Dec. 1955. Bull. Inter–Amer. trop. Tuna Comm., *8*(2):43–182.
  1966. Monthly charts of surface salinity in the eastern tropical Pacific Ocean. Bull. Inter–Amer. trop. Tuna Comm., *11*(1):1–44.

BENNETT, H. H.
  1925. Some geographic aspects of western Ecuador. Ann. Ass. Am. Geogr.. *15*.

BENNETT, WENDELL C.
  1944. Archeological regions of Colombia: a ceramic survey. Yale Univ. Press, New Haven, 120 pp.

BERRY, E. W.
  1929. Contributions to the paleontology of Colombia, Peru, and Ecuador. Johns Hopkins Univ., Stud. in Geology, (10).

BERRY, S. S.
  1959. The nature and relationship of the Panamic fauna as manifested by the Mollusca. Amer. malacol. Un., ann. Rep., for 1959: 44–45.

BLACKBURN, MAURICE
  1966. Biological oceanography of the eastern tropical Pacific: Summary of existing information. Spec. sci. Rep. U.S. Fish Wildl. Serv.–Fish., (540):18 pp.

BLACKBURN, MAURICE AND ASSOCIATES IN THE TUNA OCEANOGRAPHY RESEARCH PROGRAM (SCRIPPS)
 1962. Tuna oceanography in the eastern tropical Pacific. Spec. scient. Rep. U.S. Fish Wildl. Serv.–Fish., (400):48 pp.

BLANCHARD, DEAN HOBBS
 1962. Ecuador: crown jewel of the Andes. Vantage Press, N.Y., 228 pp.

BOCKMANN, WERNER
 1941. Unter dem Äquator. Ein südamerikanisches Erlebnisbuch. Gustav Wenzel und Sohn, Braunschweig, 192 pp.

BOEREMA, L. K. AND J. OBARRIO
 1960. The shrimp fishery of Panama. Proc. Gulf Caribb. Fish. Inst., 12th ann. Session: 10–14.

BÖHLKE, JAMES E. AND C. RICHARD ROBINS
 1968. Western Atlantic seven-spined gobies, with descriptions of ten new species and a new genus, and comments on Pacific relatives. Proc. Acad. nat. Sci. Philad., *120*(3):45–174.

BOLLAERT, WILLIAM
 1860. Antiquarian, ethnological, and other researches in New Granada, Ecuador, Peru, and Chile. Trübner & Co., London, 277 pp.

BORODIN, N. A.
 1928. Fishes. *In* Scientific results of the yacht ARA expedition during the years 1926 to 1928, while in command of William K. Vanderbilt. Bull. Vanderbilt mar. Mus., *1*(1):1–38, 5 pls., 2 charts.
 1932. Fishes. *In* Scientific results of the yacht ALVA world cruise, July 1931 –March 1932, in command of William K. Vanderbilt. Bull. Vanderbilt mar. Mus., *1*(3):65–102, 2 pls.

BOSWORTH, T. O.
 1922. Geology of the Tertiary and Quaternary periods in the northwest part of Peru. London.

BOTERO, JOSÉ MANUEL
 1950. Geografía física y de la República de Colombia. Bedout, Medellín, 222 pp.

BOULENGER, G. A.
 1887. An account of the fishes collected by Mr. C. Buckley in eastern Ecuador. Proc. zool. Soc. Lond., (1887):274–283.
 1898. Viaggio del Dott. Enrico Festa nell' Ecuador e regione vicine. Poissons de l'Équateur. Parts 1 et 2. Boll. Musei Zool. Anat. comp. R. Univ. Torino, *13*(329):1–13; *14*(335):1–8.
 1899. Viaggio del Dott. Enrico Festa nel Darien e regioni vicine. Poissons de l'Amérique centrale. Boll. Musei Zool. Anat. comp. R. Univ. Torino, *14*(346):1–4.

BRANDHORST, W.
 1958. Thermocline topography, zooplankton standing crop, and mechanisms of fertilization in the eastern tropical Pacific. J. Cons. perm. int. Explor. Mer, *24*(1):16–31.

BREDER, CHARLES M.
1925. Note on fishes from three Panama localities: Gatun Spillway, Río Tapía, and Caledonia Bay. Zoologica, N.Y., *4*:137–158.
1928. Elasmobranchii from Panama to lower California. Scientific results of the second oceanographic expedition of the PAWNEE. Bull. Bingham oceanogr. Coll., *2*(1):1–13, 12 figs.
1928. Nematognathi, Apodes, Isospondyli, Synentognathi, and Thoracostraci from Panama to lower California. Scientific results of the second oceanographic expedition of the PAWNEE. Bull. Bingham oceanogr. Coll., *2*(2):1–25, 10 figs.
1936. Heterosomata to Pediculata from Panama to lower California. Scientific results of the second oceanographic expedition of the PAWNEE. Bull. Bingham oceanogr. Coll., *2*(3):1–56, 19 figs.

BRIGGS, JOHN C.
1961. The east Pacific barrier and the distribution of marine shore fishes. Evolution, N.Y., *15*(4):545–554.

BROENKOW, WILLIAM W.
1965. The distribution of nutrients in the Costa Rica dome in the eastern tropical Pacific Ocean. Limnol. Oceanogr., *10*(1):40–52.

BRUNO, G. M.
1935. Geografía de la república del Ecuador, (2). Geografía Infantil, Libraría La Salle, Quito, 68 pp.

BUEN, F. DE
1952. Las familas de peces de importancia económica. Santiago, FAO, 311 pp.
1960. Lampreas, tiburones, rayas, y peces en la Estación de Biología Marina de Montemar, Chile. Revta. Biol. mar., *9*:3–200[1959].

BÜRGER, OTTO
1919. Reisen eines Naturforschers im tropischen Südamerika (Fahrten in Columbien und Venezuela). Dieterichsche Verlagsbuchhandlung m.b.H., Leipzig, 470 pp.

BUTLER, C. AND N. L. PEACE
1965. Spiny lobster exploration in the Pacific and Caribbean waters of the Republic of Panama. Spec. sci. Rep. U.S. Fish Wildl. Serv.—Fish., (505):26 pp.

CALDES Y TENORIO, FRANCISCO JOSÉ DE
1936. Viajes (viaje al Corazón de Barnueva). Editorial Minerva, Ministerio de Educación Nacional, Colombia, 161 pp.

CARUCCIO, A.
1886. Viaggio di circumnavigazione della R. corvetta CARACCIOLA negle anni 1881–1884. Pt. *1*:312–313.

CHAPEL, L. T.
1927. Winds and storms on the Isthmus of Panama. Mon. Wealth. Rev. U.S. Dep. Agric., *55*:519–530.

CHAPMAN, F. M.
1926. The distribution of bird-life in Ecuador. Bull. Amer. Mus. nat. Hist., *55*:1–784, Pls. 1–30, 20 text-figs.

CHUBB, L. J.
1925. The ST. GEORGE Scientific expedition. Geol. Mag., *62*:369–373.

CIEZA DE LEÓN, PEDRO DE
1864. The travels of Pedro de Cieza de León, a.d. 1532–50. Transl. and edited by Clement R. Markham, Hakluyt Society, London, 438 pp.

CIPOLLA, ARNALDO
1929. Nel Sud America del Panamá alle Ande degli Inces. Impressioni di Viaggio in Venezuela-Colombia-Panamá-Equatore-Peru. G. B. Paravia & Co., Torino, 312 pp.

CLARKE, ROBERT
1957. Migrations of marine mammals. Norsk hvalfangsttid., *46*(11):609–630.
1962. Research on marine resources in Chile, Ecuador, and Peru. Fishg. News int., *1*(5):44–50.
1962. Whale observation and whale marking off the coast of Chile in 1958 and from Ecuador towards and beyond the Galápagos Islands in 1959. Norsk hvalfangsttid., *51*(7).

CLEMENS, H. B.
1955. Fishes collected in the tropical eastern Pacific, 1952–53. Calif. Fish Game, *41*(2):161–166.

CLEMENS, H. B. AND J. C. NOWELL
1963. Fishes collected in the eastern Pacific during tuna cruises, 1952 through 1959. Calif. Fish Game, *49*(4):240–264.

COLLIER, DONALD AND JOHN V. MURRA
1943. Survey and excavations in southern Ecuador.Publ. Field Mus. nat. Hist., anthrop. Ser., *35*(528): 110 pp.

COLNETT, JAMES
1798. A voyage to the South Atlantic and round Cape Horn into the Pacific Ocean. London, 179 pp., pls., charts.

CONDAMINE, CHARLES MARIE DE LA
1778. Relation abrégée d'un voyage fait dans l'interieur de l'Amérique méridional. A. Maestricht, vxi + 379 pp.

COON, CARLETON S.
1954. The story of man. Alfred Knopf, New York.

CORLETT, E. S.
1957. Game fishing in the Gulf of Panama. Proc. 2nd int. Game Fish Conf.

CORNISH, JOHN
1955. The geography of the San Juan Delta. Unpublished master's thesis, Louisiana State University.

COVENTRY, C. A.
1944. Results of the fifth George Vanderbilt Expedition (1941). (Bahamas, Caribbean Sea, Panama, Galápagos Archipelago and Mexican Pacific islands). Monogr. Acad. nat. Sci. Philad., *6*:531–544.

CROMWELL, T.
1958. Topografía de la termoclina, corrientes horizontales y "ondulación" en

el Pacífico oriental tropical. Bull. Inter-Amer. trop. Tuna Comm., *3*(3): 153–204.

CROMWELL, T. AND E. B. BENNETT.
1959. Cartas de la deriva de superficie para el océano Pacífico oriental tropical. Bull. Inter-Amer. trop. Tuna Comm., *3*(5):235–285.

CROSSLAND, CYRIL
1927. The expedition to the South Pacific of the S. Y. ST. GEORGE. Marine ecology and coral formations in the Panama region, the Galápagos and Marquesas islands, and the atoll of Napuka. Trans. Roy. Soc. Edinburgh, *55*, pt. 2, (23):531–554.

CUERVO MARQUEY, CARLOS
1917. Orígenes etnográficos de Colombia. Proc. 2nd pan-Am. Sci. Congr., Sec. 1, Anthropol., *1*:295–329.

DAMPIER, WILLIAM AND LIONEL WAFERS
1717. Reystogten rondom de Waereldt; begrypende, in vier beknopte bockdeelen, een naauwkeurige beschryving van verscheyde nieuwe ontdekte zeen, kusten, en lande, zo in Amerika, Asia, als Afrika; . . . Stroomen, winden, havens, dieptens, engtens, landsdouwen, vrugten gewassen dieren. . . . Johannes Ratelband en Andries van Damme, Amsterdam, 394 pp. + appendix.

DANA, J. D.
1852. United States exploring expedition during the years 1838, 1839, 1840, 1841, 1842 under the command of Charles Wilkes, U.S.N. *31*, Crustacea, Pts. 1 and 2, 1618 pp.

DARWIN, CHARLES
1832–1836. Journal and natural history of the various countries visited by HMS BEAGLE. London.
1844. Geological observations in South America. London.
1959. The voyage of the BEAGLE. Harper and Bros., N.Y., 327 pp.

DAVIES, A. MORLEY
1929. Faunal migrations since the Cretaceous period. Proc. Geol. Ass., *40*.

DAWSON, E. Y.
1963. Ecological paradox of coastal Peru. Nat. Hist., N.Y., *72*(8):32–37.

DAWSON, E. Y. AND P. T. BEAUDETTE
1959. Field notes from the 1959 eastern Pacific cruise of the STELLA POLARIS. Pacif. Nat., *1*(13):1–24.

DE SYLVA, DONALD P.
1961. Game fish research in Panama, Colombia, and Ecuador. Proc. int. Game Fish Conf., (6):10 pp.
1963. Preliminary report. Station data and collection notes of the Alfred C. Glassell, Jr.—University of Miami ARGOSY Expedition to La Plata Island, Ecuador. Rept. Inst. mar. Sci., Univ. Miami, 1963, 49 pp.

DÍAZ, E.
1965. Bibliographic material on the fishes of Colombia and northwestern South America. FAO Fish. tech. Pap., (53):1–72.

DICKERSON, ROY ERNEST
1917. Climate and its influence upon the Oligocene fauna of the Pacific coast, with descriptions of some new species from the *Molopophorus luicolnensis* zone. Proc. Calif. Acad. Sci., 4th Ser., 7(6–8):157–205.

DORAN, EDWIN J.
1953.. Fish poisoning on the Río Huallaga, Peru. Tex. J. Sci., 5:204–215; 260.

DORSEY, GEORGE A.
1901. Archaeological investigations on the island of La Plata, Ecuador. Publ. Field Mus. nat. Hist., anthrop. Ser., 2(5):248–280, Pls. 43–102, Figs. 37–48.

DURHAM, J. W. AND E. C. ALLISON
1960. The geologic history of Baja California and its marine faunas. Syst. Zool., 9(2):47–91.

EGGERS, BARON H. VON
1892. Die Manglares in Ecuador. Bot. Zbl., 52:48–52.

EIGENMANN, CARL H.
1905. The fishes of Panama. Science, 22:18–20.
1910. Catalogue and bibliography of the freshwater fishes of the Americas, south of the Tropic of Cancer. Contr. zool. Lab. Univ. Indiana, 76(2):-375–511.
1912. Some results from an ichthyological reconnaissance of Colombia, South America. Part I. Indiana Univ. Stud., (8):1–27.
1913. Some results from an ichthyological reconnaissance of Colombia, South America. Part II. Indiana Univ. Stud., (18):1–31.
1920. The fishes of the rivers draining the western slope of the Cordillera Occidental of Colombia, Ríos Atrato, San Juan, Dagua, and Patía. Indiana Univ. Stud., 7(46): 1–19.
1921a. The origin and distribution of the genera of fishes of South America west of the Maracaibo, Orinoco, Amazon and Titicaca basins. Proc. Am. phil. Soc., 60(1):1–6.
1921b. The nature and origin of the fishes of the Pacific slope of Ecuador, Peru, and Chile. Proc. Am. phil. Soc., 60(4):503–523.
1922. The fishes of Northwestern South America. Part I. Mem. Carneg. Mus., 9:218 pp.
1923. The fishes of the Pacific slope of South America and the bearing of their distribution on the history of the development of the topography of Peru, Ecuador, and western Colombia. Am. Nat., 57:193–210.

EIGENMANN, CARL H. AND W. R. ALLEN
1942. Fishes of western South America. Univ. Kentucky, Lexington, 494 pp.

EIGENMANN, CARL H. AND ARTHUR HENN
1914. On new species of fishes from Colombia, Ecuador, and Brazil. Indiana Univ. Stud., 2(24):231–234.

EIGENMANN, CARL H., ARTHUR HENN, AND CHARLES WILSON
1914. New fishes from western Colombia, Ecuador, and Peru. Indiana Univ. Stud., (19):1–15.

EISENMANN, R.
1958. The shrimp industry of Panama. Proc. Gulf Caribb. Fish. Inst., 10th Sess.; 34–36.

EKMAN, SVEN
1953. Zoogeography of the sea. Sidgwick and Jackson, Ltd., London, xiv + 412 pp.

ELLIS, ROBERT W.
1961. Informe al Gobierno del Ecuador sobre la biología del camaron. FAO/ETAP, Rome.

ENOCK, CHARLES REGINALD
1914. Ecuador: its ancient and modern history, topography and natural resources, industries and social development. T. F. Unwin, London, 375 pp.

EVERMANN, B. W. AND E. L. GOLDSBOROUGH
1909. Notes on some fishes from the Canal Zone. Proc. biol. Soc. Wash., *22*:95–104.

FARRINGTON, KIP, JR.
1942. Pacific game fishing. Coward-McCann, N.Y., xii + 277.
1953. Fishing the Pacific. New York, Coward-McCann, Inc., xviii + 297 pp.

FAXON, W.
1893. Reports on dredging operations off the west coast of Central America to the Galápagos, to the west coast of Mexico, and in the Gulf of California, in charge of Alexander Agassiz, carried on by the U.S. Fish Commission steamer ALBATROSS during 1891. Bull. Mus. comp. Zool. Harv., *24*:149–220.

FERDON, EDWIN N., JR.
1940–1941. Reconnaissance in Esmeraldas. Palacio, *47*:257–272 (1940); *48*:7–15 (1941).
1950. Studies in Ecuadorian geography. Monogr. School Amer. Res., *15*, Santa Fe, New Mexico.

FERMÍN CEBALLOS, PEDRO
1888. Geografía de la República del Ecuador. Imprenta del Estado, Lima, 371 pp.

FESTA, ENRICO
1901. Nel Darien e nel l'Ecuador; diario di viaggio di un naturalista Torino. Unione tip., editrice Torinese, xvi + 397 pp.

FINNEGAN, S.
1931. Report on the Brachyura collected in Central America, the Gorgona and Galápagos islands, by Dr. Crossland on the ST. GEORGE expedition to the Pacific, 1924–1925. J. Linn. Soc., *37*:607–673, text figs. 1–6.

FISCHER, W.
1966. El desarrollo pesquero en América Latina en su aspecto biológico. FAO Fish. Circ., 102:59 pp.

FISHER, R. L.
1958. Preliminary report on expedition Downwind, Univ. of California,

Scripps Inst. Oceanogr., IGY cruise to the southeast Pacific IGY, gen. Rep., (2).

FLEMING, R. H.
1938. Tides and tidal currents in the Gulf of Panama. J. mar. Res., *1*:192–206.
1940. A contribution to the oceanography of the Central American region. Proc. Sixth Pacif. Sci. Congr., *3*:167–175.

FOOD AND AGRICULTURAL ORGANIZATION, UNITED NATIONS
1960. Informe al Gobierno de la República de Panama, sobre los recursos camaroneros panameños. Basado en el trabajo de L. K. Boerema, Biólogo pesquero marino de la FAO/EPTA. Rep. FAO/EPTA, (1537):45 pp.

FORSBERGH, E. D.
1963. Some relationships of meteorological, hydrographic, and biological variables in the Gulf of Panama. Bull. Inter-Amer. trop. Tuna Comm., *7*(1):1–109.

FORSBERGH, E. D. AND WILLIAM N. BROENKOW
1965. Oceanographic observations form the eastern Pacific Ocean collected by the R/V SHOYO MARU, October 1963–March 1964. Bull. Inter-Amer. trop. Tuna Comm., *10*(2):85–159.

FORSBERGH, E. D. AND J. JOSEPH
1964. Biological production in the eastern Pacific Ocean. Bull. Inter-Amer. trop. Tuna Comm., *8*(9):477–527.

FOWLER, HENRY W.
1938. The fishes of the "George Vanderbilt South Pacific Expedition," 1937. Monogr. Acad. nat. Sci. Philad., (2):viii + 349 pp., 12 pls.
1939. Fishes from the Pacific slope of Colombia, Ecuador, and Peru. Notul. Nat., (33):7 pp.
1942a. Lista de peces de Colombia. (Introduction by C. Miles). Revta. Acad. colomb. Cienc. exact. fís. nat., 5(17):128–138.
1942b. Los peces del Perú. Catálogo sistemático de los peces que habitan en aguas peruanas (continuación). Boln. Mus. Hist. nat. Javier Prado, *6*:352–381.
1943. Lista de peces de Colombia. Revta. Acad. colomb. Cienc. exact. fís. nat., *5*(17):128–138.
1944. Results of the Fifth George Vanderbilt Expedition (1941). Monogr. Acad. nat. Sci., Philad., (6):57–529 (western Caribbean) + 475–529 (eastern Pacific).
1945a. Colombian Zoological Survey. Pt. 1—The fresh-water fishes obtained in 1945. Proc. Acad. nat. Sci. Philad., *97*:93–135.
1945b. Los peces del Perú. Catálogo sistemático de los peces que habitan en aguas peruanas. Univ.nac. de San Marcos, Lima.
1945c. Los peces del Perú—Catálogo sistemático de los peces que habitan en aguas peruanas. Boln. Mus. Hist. nat. Javier Prado, (1945):298 pp.
1950. Colombian zoological survey. Part VI. The fishes obtained at Totumo, Colombia, with descriptions of two new species. Notul. nat., (222):108.

FRANZE, BRUNO
1927. Die Niederschlagsverhältnisse in Südamerika. Petermanns geogr. Mitt., Ergänzungsband, (42), Ergänzungsheft, (193) for 1927: 79 pp., 1 pl.

FRASER, C. MCLEAN
1943. General account of the scientific work of the VELERO III in the eastern Pacific, 1931–1941, part 3. A ten-year list of the VELERO III collecting stations. Allan Hancock Pacific Exped., *1*(1,2,3):259–431, charts 1–115.

FÜHRMANN, OTTO AND EUGEN MAHOR
1914. Voyage de Exploration Scientifique en Colombie. Poteries anciennes de la Colombie. Por Th. Delachoux de la Société Neuchâteloise de Sciences Naturelles. Vol. 5. Attinger Fréres, Neuchâtel, 116 pp.

GANSSER, A.
1950. Geological and petrographical notes on Gorgona Island in relation to northwestern South America. Schweiz. miner. petrogr. Mitt., *30*:219–237.

GARMAN, SAMUEL W.
1877. Notes on some fishes from the western coast of South America. Proc. Boston Soc. nat. Hist., 1875–76, *18*:202–205.
1899. Reports on an exploration of the west coast of México, Central and South America, and off the Galápagos Islands in charge of Alexander Agassiz, by the U.S. Fish Commission steamer ALBATROSS, during 1891. The fishes. Mem. Mus. comp. Zool. Harv., *(24)*:431 pp., 97 pls.
1906. Pisces. *In* Vertebrata from the Savanna of Panama. No. 12. Bull. Mus. comp. Zool. Harv., *46*:209–230.
1913. The Plagiostomia (sharks, skates and rays). Mem. Mus. comp. Zool. Harv., *36*:515 pp.

GARTH, JOHN S.
1948. The Brachyura of the ASKOY expedition with remarks on carcinological collecting in the Panama Bight. Bull. Amer. Mus. nat. Hist., *92*(1):1–66, 8 pls., 5 text-figs.

GIERLOFF-EMDEN, H. G.
1959. Die Küste von El Salvador. Eine morphologisch-oceanographische Monographie. Acta humboldt., *2*

GIGLIOGLI, ENRICO H.
1875. Viaggio intorno al globo della r. pirocorvetto MAGENTA negli anni 1865–66–67–68 sotto il commando del capitano di fregata V. F. Arminjon. U. Maisner, Milano, xxxviii + 1031 pp.

GILBERT, C. H.
1890. A supplementary list of fishes collected at the Galápagos Islands and Panama, with descriptions of one new genus and three new species. Proc. U.S. natn. Mus., *13*:449–455.
1891. A supplementary list of fishes collected at the Galápagos Islands and Panama, with descriptions of one new genus and three new species. Scientific Results of Explorations . . . ALBATROSS, No. 19. Proc. U.S. natn. Mus., 1890 (1891), *13*:449–455.

GILBERT, CHARLES H. AND EDWIN C. STARKS
1904. The fishes of Panama Bay. Mem. Calif. Acad. Sci., *4:*1–304, 33 pls., 62 figs.

GILL, T. N.
1864. Descriptive enumeration of a collection of fishes from the western coast of Central America, presented to the Smithsonian Institution by Captain John M. Dow. Proc. Acad. nat. Sci. Philad., 1863 (1864): 162–180.

GILMARTIN, MALVERN
1964. Compilación bibliográfica sobre la oceanografía de las aguas litorales de Colombia, Ecuador, y Perú, con expecial referencia al fenómena "El Niño." Boln. cient. Inst. nac. Pesca, Ecuador, *1*(1):15 pp.

GILMORE, R. M.
1950. Fauna and ethnozoology of South America. Bull. Bur. Am. Ethnol., *6*(143):345–464.

GOEZ, RAMÓN CARLOS
1947. Geografía de Colombia. Fondo de Cultura Económica, Pánuco, México, 214 pp., pls.

GRANJA, JULIO C.
1957. Bosquejo de la geología del Ecuador. Editorial Universitaria, Quito, 50 pp.

GREENBACK, J. T.
1960. Informe sobre el centro latinoamericano de capacitación en piscicultura y pesca continental dessarrollado bajo los auspicios del gobierno de Colombia y de la organización de las naciones unidas para la agricultura y la alimentación en Buga, Colombia, 1 febrero–12 marzo 1960. Rep. FAO/ETAP, (1262): 22 p., mimeo.

GREGORY, J. W.
1930. The geological history of the Pacific Ocean. Q. Jl. geol. Soc. Lond., *86:*ix–xxi.

GÜNTHER, A. C. L. G.
1861. On a collection of fishes sent by Capt. Dow from the Pacific coast of Central America. Proc. zool. Soc. Lond., 1861: 370–376, ill.
1869. An account of the fishes of the States of Central America, based on collections made by Capt. J. M. Dow, F. Goodman, Esq., and O. Salvin, Esq. Trans. zool. Soc. Lond., *6*(7): 377–494, 24 pls.

GUNTHER, E. R.
1936. A report on oceanographical investigations in the Peru coastal current. Discovery Rep., *13*:107–276.

HAGBERG, ANDERS H., ROBERT W. ELLIS, AND MALVERN GILMARTIN
1967. Compilación bibliográfica sobre pesquerías en Centro América y Panamá (oceanografía, biología marina, limnologia, e industria). Bol técn., Proy. reg. Desarrollo Pesq. Centroamerica, *1*(1):57 pp.

HALSTEAD, BRUCE W.
1953. Report on an investigation of poisonous and venomous fishes at Cocos, Galápagos, and La Plata islands during 4 December 1952 to 28 January

1953. Loma Linda, Calif., Coll. med. Evang., Office of Nav. Res., Biol. Branch., Contract no. Nonr-205(00), 14 pp.

HEACOCK, J. G. AND J. L. WORZEL
1955. Submarine topography west of Mexico and Central America. Bull. geol. Soc. Am., *66*:773–776.

HENN, A. W.
1914. Indiana University expeditions to northwestern South America. Science, N.Y., *40:*602–606.

HENRY, A. J.
1922. Rainfall of Colombia. Mon. Weath. Rev. U.S. Dep. Agric., *50:*189–190; 262.

HERRE, A. W.
1936. Fishes of the Crane Pacific expedition. Publ. Field Mus. nat. Hist., zool. Ser., *21*(353):472 pp.

HERTLEIN, L. G. AND A. M. STRONG
1955. Marine molluscs collected during the ASKOY expedition to Panama, Colombia, and Ecuador in 1941. Bull. Amer. Mus. nat. Hist., *107*:159–318.

HETHERINGTON, H., PUBL.
1846. Tropical information. A treatise on the history, climate, soil, production, manufactures, commerce . . . of Venezuela, with like notices of New Granada and Ecuador, and a slight glance at Bolivia and Peru. R. Johnson, London, 192 pp.

HILDEBRAND, SAMUEL F.
1938. A new catalogue of the fresh-water fishes of Panama. Publ. Field Mus. nat. Hist., zool. Ser., *22* (4):88 pp.
1939. The Panama Canal as a passageway for fishes, with lists and remarks on the fishes and invertebrates observed. Zoologica, N.Y., *24:*15–45, Pls. 1,2.
1946. A descriptive catalogue of the shore fishes of Peru. Bull. U.S. natn. Mus., (189): xi + 530 pp.

HILL, ROBERT T.
1898. The geological history of the isthmus of Panama and portions of Costa Rica. Bull. Mus. comp. Zool. Harv., *28*(5):151–285, 19 pls.

HOEDEMAN, J. J.
1960. A list of type specimens of fishes in the Zoological Museum, University of Amsterdam. 1. Order Mugiliformes. Beaufortia, *7*(87):211–217.

HOLDRIGE, L. R., *et al.*
1947. The forests of western and central Ecuador. Forest Service, U.S. Dept. Agric., Wash., D.C., 263 pp.

HOLMES, R. W. AND M. BLACKBURN
1960. Physical, chemical, and biological observations in the eastern tropical Pacific Ocean: SCOT expedition, April–June 1958. Spec. sci. Rep. U.S. Fish Wildl. Serv.—Fish., (345):106 pp.

HOLMES, R. W. AND OTHER MEMBERS OF THE SCRIPPS COOPERATIVE OCEANIC
PRODUCTIVITY EXPEDITION
  1958. Physical, chemical, and biological oceanographic observations obtained
        on expedition SCOPE in the eastern tropical Pacific, Nov.–Dec. 1956.
        Spec. sci. Rep. U.S. Fish Wildl. Serv—Fish., (279):117 pp.

HOWARD, G. V. AND E. GODFREY
  1951. A summary of the information on the fisheries and fisheries resources of
        Latin America. Food and Agricultural Organization, 262 p. mimeo.

HOWARD, JOHN K. AND SHOJI UEYANAGI
  1965. Distribution and relative abundance of billfishes (Istiophoridae) of the
        Pacific Ocean. Stud. trop. Oceanogr. Miami, *2*:1–134, 37 figs. in atlas.

HUBBS, CARL L.
  1953. Geographic and systematic status of the fishes described by Kner and
        Steindachner in 1863 and 1865 from fresh waters in Panama and
        Ecuador. Copeia, 1953 (3):141–148.
  1959. Initial discoveries of fish faunas on sea mounts and offshore banks in the
        eastern Pacific. Pacif. Sci., *13*(4):311–315.

HUBBS, CARL L. AND G. I. RODEN
  1964. Oceanography and marine life along the Pacific coast of middle Amer-
        ica. Handbook of Middle American Indians, vol. 1. R. C. West, Univ.
        of Texas, Austin.

HUBBS, CARL L. AND RICHARD H. ROSENBLATT
  1961. Effects of the equatorial currents of the Pacific on the distribution of
        fishes and other marine animals. Abstr. Symp. Pap., 10th Pac. Sci.
        Congr., Honolulu: 340–341.

HUMBOLDT, FRIEDRICH H. ALEXANDER VON
  1805. Voyage aux régions equinoxiales du nouveau continent, fait en 1799–
        1804, etc. Paris, 24 vols., 1805–1837, pls. and atlases.
  1808. Aussichten der Natur mit wissenschaftlichen Erläuterungen. J. G. Cotla,
        Tübingen, viii + 334 pp.
  1824. Essai politique sur le royaume de la nouvelle-Espagne. 5 vols., 2nd ed.,
        Paris.

HUNTER, J. R. AND C. T. MITCHELL
  1966. Association of fishes with flotsam in the offshore waters of Central
        America. Fishery Bull. Fish Wildl. Serv. U.S., *66*(1):13–29.

IDDINGS, A. AND A. A. OLSSON
  1928. Geology of north-west Peru. Bull. Am. Ass. Petr. Geol., for 1928: 149.

INSTITUTE OF MARINE RESOURCES
  1965. Review of the coastal fisheries of the west coast of Latin America. Univ.
        California in collaboration with Inter-American tropical Tuna Com-
        mission. Inst. Marine Resources Ref., (65-4):1–152.

JAMES, PRESTON E.
  1942. Latin America. Odyssey Press, N.Y., vxiii + 908 pp.

JANSEN, ARNOLD WILTON
  1944. La pesca en la costa del Pacífico. Colombia, *1*(3,4):99–107; 263.

JENNINGS, A. H.
  1950. World's greatest observed point rainfalls. Mon. Wealth. Rev. U.S. Dep.
         Agric., *78*:4–5.

JONES, STEWART M.
  1950. Geology of Gatun Lake and vicinity, Panama. Bull. geol. Soc. Am.,
         61:893–921.

JORDAN, DAVID S.
  1886. A list of the fishes known from the Pacific coast of tropical America,
         from the Tropic of Cancer to Panama. Proc. U.S. natn. Mus., 1885
         (1886), *8*(24–25):361–394.

JORDAN, DAVID S. AND C. H. BOLLMAN
  1890. Description of new species of fishes collected at the Galápagos Islands
         and along the coast of the United States of Colombia, 1877–1888. Proc.
         U.S. natn. Mus., 1889 (1890), *12*(770):149–183.

JORDAN, DAVID S. AND B. W. EVERMANN
  1896–1900. The fishes of North and Middle America. Bull. U.S. natn. Mus.,
         pt. 1, 1896, pp. I–IX, 1–1240; pt. 2, 1898, pp. I–XXX, 1241–2183; pt.
         3, pp. I–XXIV, 2184–3136; pt. 4, pp. I–CI, 3157–3313; 292 pls.

JORDAN, DAVID S., B. W. EVERMANN, AND C. H. WALTON
  1882. Descriptions of nineteen new species of fishes from the bay of Panama.
         Bull. U.S. Fish.Comm., *1*:306–318.

JORDAN, DAVID S., B. W. EVERMANN, AND H. W. CLARK
  1930. Check list of the fishes and fishlike vertebrates of North and middle
         America north of the northern boundary of Venezuela and Colombia.
         Rep. U.S. Commn. Fish., *1928*(2):670 pp.

JORDAN, DAVID S. AND CHARLES H. GILBERT
  1882. Description of nineteen new species of fishes from the bay of Panama.
         Bull. U.S. Fish Commn., 1881 (1882), *1*:306–335.

KARSTEN, HERMANN
  1858. Beiträge zur Geologie des westlichen Columbien. Amtl. Bericht wiener
         Naturforscherversammlung.
  1886. Géologie de l'ancienne colombie bolivarienne Vénézuela, Nouvelle-
         Grenade, et Ecuador. R. Friedländer & Sohn, Berlin, 62 pp.

KENDALL, W. C. AND L. RADCLIFFE
  1912. Report on the scientific results of the expedition of the eastern tropical
         Pacific in the charge of A. Agassiz, by the U.S. Fish Commission steamer
         ALBATROSS, from Oct. 1904 to Mar. 1905, Commander L. M. Carret,
         U.S.N. commanding. 25. The shore fishes. Mem. Mus. comp. Zool.
         Harv., *35*(3):75–171, 8 pls

KIRKPATRICK, R. Z.
  1926. Panama tides. Proc. U.S. nav. Inst., *52*(4):660–664.

KLAWE, W. L.
  1964. Food of the black-and-yellow sea snake, *Pelamis platurus*, from Ecua-
         dorian coastal waters. Copiea, 1954(4):712–713.

KNER, RUDOLPH AND FRANZ STEINDACHNER
1865. Neue Gattungen und Arten von Fischen aus Central-Amerika gesammelt von Prof. Moritz Wagner. Abh. bayer. Akad. Wiss. Math.-phys. Kl., 1864 (1865), *10:*1–61, 6 pls.

KNOCH, KARL
1930. Klimakunde von Südamerika, Bd. 2, Teil G., Berlin.

KNOCHE, WALTER
1931–1932. Bio- und medizinsch-geographische Studien auf einer Reise nach Ecuador. Phönix. 2 Verh. dt. wiss. Ver., Buenos Aires.
1932. Klimatische Beobachtungen auf einer Reise in Ecuador. Verh. dt. wiss. Ver. Santiago, Chile. N.F., Bd. 2. Walter Gnadt, Santiago de Chile, 46 pp.

KOEPCKE, HANS-WILHELM
1951. Clave para identificar los peces communes de la costa peruana. Serie Divulgs. cient. No. 1, Minist. Agric., Dir. Pesquería y Caza, Lima, 68 pp.
1959. Beiträge zur Kenntnis der Fische Perus. II. Beitr. neotrop. Fauna, *1:*249–268.

LARREA, CARLOS MANUEL
1924. Geographical notes on Esmeraldas, northwestern Ecuador. Geogr. Rev., *14:*373–387, Figs. 1–10.
1948–1952. Bibliografía científica del Ecuador. Suppl. Vol. 2, 4, 5, to Boletín de Informaciones Científicos Nacionales.
1952. Bibliografía científica del Ecuador. Ediciones Cultura Hispánica, Madrid, 492 pp.
1960. El archipiélago de Colón (Galápagos) descubrimiento, exploraciones científicas y bibliografía de las islas. Ed. 2., Editorial Casa de la Cultura Ecuatoriana, Quito, 423 pp.; bibl.:263–379.

LASSO, RAPHAEL V.
1944. The wonderland Ecuador. Alpha-Ecuador Publications, New York, xiv + 297 pp.

LÉVINE, V.
1914a. Colombia. D. Appleton & Co., N.Y., 220 pp.
1914b. Colombia. Sir Isaac Pitman, Ltd., London, xii + 220 pp.

LI, C. G.
1930. The Miocene and Recent molluscs of Panama Bay. Bull. geol. Soc. Am., *9:*249–296, 8 pls.

LINDNER, M. J.
1957. Survey of shrimp fisheries of Central and South America. Spec. sci. Rep. U.S. Fish Wildl. Serv.—Fish., (325):166 pp.

LINKE, LILO
1954. Ecuador: country of contrasts. Royal Inst. Int. Affairs, London and New York, ix + 173 pp.
1960. Ecuador, country of contrasts. London, New York, Royal Inst. of Internat. Affairs, 3rd ed., ix + 193 pp.

LLERAS CODAZZI, RICARDO
1926. Notas geográficas y geológicas. Imprenta Nacional, Bogotá, 125 pp.

LOESCH, HAROLD AND QUINTO AVILA
1964. Claves para identificación de camarones peneidos de interés comercial en el Ecuador. Bol cient. técn. Inst. nac. Pesca Ecuador, *1*(2).

LÓPEZ, FELICÍSIMO
1907. Atlas geográfico del Ecuador, arreglado según la carta del Dr. Teodoro Wolf.

MAACK, G. A.
1874. Report on the geology and natural history of the isthmus of Chocó, of Darien, and of Panama, 1872. House Misc. Doc. 113 (Dated Cambridge, Mass., Oct. 1, 1871).

MACDONALD, DONALD F.
1919. The sedimentary formations of the Panama Canal Zone, with special reference to the stratigraphic relations of the fossiliferous beds. Bull. U.S. natn. Mus., (103):125–145; 173.

MARMER, H. A.
1930. Panama tides. Proc. U.S. nav. Inst., *56*:1003–1008.

MAULL, OTTO
1937. Nature, culture, and science in South America. Ecuador, *In* Handbuch der geographische Wissenchaft, *5*.

MEEK, S. E. AND S. F. HILDEBRAND
1923–1928. The marine fishes of Panama. Publ. Field Mus. nat. Hist., zool. Ser., *15*(1–3); 1923, (1): *215:*v–xi + 1–330, 24 pls: 1925, (2):226; xv–xix + 345–708, Pls. 25–71; 1928, (3): *249:* xxv–xxxi + 709–1045, Pls. 72–102.

MENARD, H. W.
1960. The East Pacific Rise. Science, N.Y., *132:*1737–1746.

MENDOZO NIETO, JORGÉ
1942. Geografía ilustrada del Chocó. Bogotá.

MEREDITH, D.
1939. Voyages of the VELERO III, a pictorial vision with historical background through tropical seas to equatorial lands aboard M/V VELERO III. Los Angeles, 286 pp., pls.

MERIZALDE DEL CARMEN, P. BERNARDO
1921. Estudio de la costa colombiana del Pacífico. Estado Mayor General, Bogotá, 248 pp., illus.

MILES, C.
1942. Importancia de la ictiología en Colombia. Caldasía, (5):51–54.

MILNE-EDWARDS, A.
1868–1897. Mission scientifique au Mexique et dans l'Amérique Centrale. Ouvrage publiée par les soins du ministre de l'instructions publique. Impr. Impériale, Paris, 10 vols., illus., pls.
1875–1880. Études sur les xiphosures et les crustacés de la région Mexicaine.

Mission Scientifique au Mexique et dans l'Amérique Centrale. Pt. 5, 368 pp.

MORCH, O. A. L.
1859–1861. Beiträge zur Molluskenfauna Central-Americas. Malakozool. Bl., *6:*102–126 (1860); 193–213 (1861).

MOSQUERA, TOMÁS CIPRIANO DE
1953. Memoir on the physical and political geography of New Granada. Transl. from the Spanish by Theodore Dwight. T. Dwight, N.Y., 105 pp.

MÜLLER, RICHARD F.
1935. Ekuador. Die wundervolle Werkstatt der Natur. Editorial Jouvin, Guayaquil, 16 pp.
1936. Scientific information for travelers in Ecuador. Land forms. 2 ed., Editorial Jouvin, Guayaquil, 62 pp.

MURPHY, GRACE E. B.
1943. Flight to Ecuador. Nat. Hist., N.Y., *52:*62–69, 19 illus.

MURPHY, ROBERT C.
1925a. Equatorial vignettes. Impressions of the coasts of Ecuador and Peru. Nat. Hist., N.Y., *25*(5):431–449.
1925b. Oceanic and climatic phenomena along the west coast of South America during 1925. Geogrl. Rev., *16:*26–54.
1936. Oceanic birds of South America. American Museum of Natural History, New York, 2 vols., 1245 pp.
1939a. The littoral of Pacific Colombia and Ecuador. Geogrl. Rev., *29:*1–33.
1939b. Racial succession in the Colombian Chocó. Geogrl. Rev., *29:*461–471.
1941. The ASKOY expedition of the American Museum of Natural History to the eastern tropical Pacific. Science, N.Y., *94:*57–58.
1942. Pacific campaign of the schooner ASKOY. Darien, Colombia, Ecuador. Trans. Amer. geophys. Un., 23rd Ann. Meet. Part II; 336–338.
1944a. Beyond the continental shelf. Nat. Hist., N.Y., *53*(2):303–309.
1944b. To the Chocó in the schooner ASKOY. Nat. Hist., N.Y., *53*(5):200–208.
1944c. Among the Pearl Islands. Nat. Hist., N.Y., *53*(6):274–281.
1944d. Wet lands and dry seas. Nat. Hist., N.Y., *53*(8):350–356.
1944e. Mountain and sea in the Chocó. Nat. Hist., N.Y., *53*(10):474–481.
1945. Island contrasts. Nat. Hist., N.Y., *54:*14–23.

MURRAY, JOHN
1895. Report of the scientific results of the voyage of HMS CHALLENGER during the years 1872–1876. Rep. Challenger Expedition, a summary of the scientific results. London, Edinburgh, and Dublin, Part 2: xix + 797–1608.

MYERS, G. S.
1941. The fish fauna of the Pacific Ocean, with special reference to zoogeographical regions and distribution as they affect the international aspects of the fisheries. Proc. Sixth Pacif. Sci. Congr., 1940, *3:*201–210.

MYERS, G. S. AND C. B. WADE
1946. New fishes of the families Dactyloscopidae, Microdesmidae, and Anten-

154 Bibliography

nariidae from the west coast of Mexico and the Galápagos Islands with a brief account of the use of rotenone fish poison in ichthyological collecting. Allan Hancock Pacif. Exped., *9*(6):151–179.

NEALE, E. ST. J.
1866. On the discovery of gold deposits in the district of Esmeraldas, Ecuador. Q. Jl. geol. Soc. Lond., *22*.

NICHOLS, J. T. AND R. C. MURPHY
1944. A collection of fishes from the Panamá Bight, Pacific Ocean. Bull. Amer. Mus. nat. Hist., *83*(4):217–260, text-figs. 1–6.

NOBILI, G.
1901. Viaggio del Dr. Enrico Festa nella Republica dell Ecuador e regione vicine. Decapodi e Stomatopodi. Boll. Musei Zool. Anat. comp. R. Univ. Torino, *16*(415):1–58.

NYGREN, N. E.
1950. The Bolivar geosyncline of northwestern South America. Bull. Am. Ass. Petr. Geol., *34:*1998–2006.

OLSSON, A. A.
1931. The Oligocene of Ecuador. Bull. Am. Paleont., *17*(63):179.
1961. Mollusks of the tropical eastern Pacific, particularly from the southern half of the Panamic-Pacific faunal province (Panama to Peru). Paleontol. Res. Inst., Ithaca, N.Y., 572 pp.
1964. Neogene mollusks from northwestern Ecuador. Paleontological Research Inst., Ithaca, N.Y., 258 pp., 38 pls.

ONFFROY DE THORON, DON ENRIQUE
1866. Amérique équatoriale; son historie pittoresque et politique, sa géographie et ses richesses naturelles. Jules Renouard, Ed., Paris, xii + 468.

OPPENHEIM, VICTOR
1949. Geología de la costa sur del Pacífico de Colombia. Boln. Inst. geofís. Andes Colomb., *1:*1–23.
1952. The structure of Colombia. Trans. Am. geophys. Un., *30:*739–748.

Orces, Gustavo.
1942. Los ofídios venenosos del Ecuador. Flora, Quito, *2*(5–6): 147–155; *3*(7–10):165–170.
1947. Algunos géneros de peces no señalados previamente en el Ecuador. Boln. Infs. cient. nac., Quito, *1*(4):33–37.
1950. Sobre una colección de peces marinos obtenida en el noroeste del Ecuador. Boln. Informs. cient. Nac. Pesca, Quito, Pt. 1, *3*(35):294–320.
1951. Sobre una collección de peces marinos obtenida en el noroeste del Ecuador. Boln. Informs. cient. Nac. Pesca, Quito, Pt. 2, *4*(43):353–369.
1955. Observaciones sobre los Elasmobranquios del Ecuador. Revta. Biol. mar., *5*(1–3):85–110.
1959a. Contribuciones al conocimiento de los peces marinos del Ecuador. An. Univ. Centr. Ecuador, *88:*107–160.
1959b. Nombres vulgares y su equivalenta científico de peces marinos de las costas del Ecuador. Cienc. y Nat., Quito, *2:*15–19, 1 fig.

1959c. Peces marinos del Ecuador que se conservan en las colecciones de Quito. Cienc. Nat., *2:*72–91.

ORTÍZ, SERGIO ELIAS
1937. Contribución a la bibliografía sobre ciencias etnológicas de Colombia. (Idearium, Suplemento No. 1, Pasto). Imprenta del Departamiento, Pasto, 66 pp.

ORTIZ BORDA, L.
1961. Importancia de una facultad de ciencias del mar. Hechos y Notic., Bogotá, (8):14–15.

OSBURN, R. C. AND J. T. NICHOLS
1916. Shore fishes collected by the ALBATROSS expedition in lower California with description of new species. Bull. Am. Mus. nat. Hist., *35:*139–181, 15 figs.

OSORIO-TAFALL, B. F.
1951. Better utilization of fisheries resources in Latin America. Fish. Bull. F.A.O., *4*(3):2–25.

OSWALD, ERLING
1964. FAO/UN Report to the Government of Ecuador on improvement of small coastal fishing craft, based on the work of Erling Oswald. FAO/EPTA Fisheries Expert (Spanish). FAO/UN Report, (1857), Rome, tabl., illus., graphs.

PARISEAU, EARL J.
1963. Handbook of Latin American Studies. Univ. Florida Press, Gainesville, Florida, 25 vols.

PAZ ANDRADE, VALENTIN
1956. Informe al Gobierno de Colombia sobre el fomento de la industria pesquera. Rep. FAO/ETAP, (509):1–13.

PAZ Y MIÑO, L. T.
1950. Bibliografía geografía ecuatoriana. Quito.

PÉREZ ARBELÁEZ, E.
1953. Recursos naturales de Colombia. Su génesis, su medida, su aprovachamiento, conservación, y renovación. Dificultades naturales de Colombia y lucha contra ellas. Capítulo I. La. posición continental, el mar. Instituto Geográfico de Colombia "Agustín Codazzi," Bogotá, 96 pp.

PEREZ DE BARRADAS, JOSÉ
1954. Orfebrería prehispánica de Colombia, estilo calima; obtra basada en el estudio de las collecciones del Museo del Oro del Banco de la República, Bogotá. Madrid, 2 vols., 366 pp. text + pls.

PESTA, OTTO
1931. Ergebnisse der Österreichischen biologischen Costa-Rica Expedition 1930. I. Teil. Crustacea Decapoda aus Costa-Rica. Annln. naturh. Mus. Wien, *45:*173–181.

PLATT, RAYE ROBERTS
1956. Ecuador. Doubleday Inc., Garden City, N.Y., 64 pp.

1959. Colombia. (Prepared with the cooperation of the American Geographical Society.) Doubleday Inc., Garden City, N.Y., 64 pp.

PORTIG, W. H.
1965. Central American rainfalls. Geogrl. Rev., *55*(1):68–90.

POSADA Y ARANGA, ANDRES
1909. Estudios científicos, con algunos otros escritos sobre diversos temas. Contr. Estudio Fauna Colombiana, Medellín, 432 pp. Los peces: 285–322.

PREUSS, KONRAD TH.
1914. Reisebrief aus Kolumbien. Z. für Ethnol., *46:*106–113.
1929. Monumentale vorgeschichtliche Kunst, Ausgrabungen im Quellgebiet des Magdelena [sic] in Kolumbien und ihre Ausstrahlungen in Amerika. Vandenhoeck and Ruprecht, Göttingen, 116 pp.

QUIROGA, DOMINGO
1964. Apuntes e informaciones sobre las pesquerías en la provincias del Guayas y Los Ríos. Boln. Informs. Inst. nac. Pesca Ecuador, *1*(4):84 pp., maps.

QUIROGA, DOMINGO AND ANÍBAL ORBES ARMAS
1963a. Apuntes e informaciones sobre la situación de la producción pesquera Ecuatoriana y sus mercados. Boln. Informs. Inst. nac. Pesca Ecuador, *1*(3):1–24.
1964b. Apuntes e informaciones sobre las pesquerías en la Provincia de Esmeraldas. Boln. Informs. Inst. nac. Pesca Ecuador, *1*(6):22 pp., illus., maps.
1964c. Apuntes e informaciones sobre las pesquerías en el Archipiélago de Colón (Islas Galápagos). Boln. Informs. Inst. nac. Pesca Ecuador. *1*(5).

QUIROGA-RÍOS, D. A.
1958. El desarrollo de las pesquerías ecuatorianas. Rep. FAO/EPTA, (745).

RECASENS, JOSÉ DE AND VICTOR OPPENHEIM
1943–1944. Análisis tipológico de matenates cerámicos y líticos, procedentes del Chocó. Revta. Inst. etnol. nac., Bogotá, *1:*331–409.

REGAN, C. TATE
1906–1908. Pisces. *In* Godman, F. D. and O. Salvin, Biologia Centrali-Americana, London, xxxiii + 203 pp., 7 maps, 1 fig., 26 pls.

RENDAHL, H.
1941. Fische aus dem pazifischen Abflussgebiet Kolumbiens. Ark. Zool., *33A*(4):15 pp.

REYES, OSCAR EFRÉN
1956. Breve historia del Ecuador. Editorial "Fray fodoco Ricke," Quito, *1–3* in 2 vols.

RICKER, K. E.
1959. Fishes collected from the Revillagigedo Islands during the 1954–1958 cruises of the MARIJEAN. Inst. Fish. Univ. Brit. Columb., Mus. Contr., (4):1–10.

RIOJA, E.
1962. Caracteres biogeográficos de México y de Centro América. Revta. Soc. mex. Hist. nat., *23:*27–50.

RIPLEY, WILLIAM ELLIS
1964. Survey of the dominant conditions affecting the development of the Cartagena (Colombia) Fishery. Market News Leafl., Wash., (87):58 pp.

ROBINS, C. RICHARD
1957. The Charles F. Johnson Oceanic Gamefish Investigations Progress Rep. No. 4. Unv. Miami Mar. Lab., Rep. 57–28, 10 pp. (Mimeogr.)
1958. Observations on oceanic birds in the Gulf of Panama. Condor, *60*(5): 300–302.

ROBINSON, WIRT
1895. A flying trip to the tropics. Riverside Press, Cambridge, x + 194 pp.

RODEN, G. I.
1962. Oceanographic aspects of the eastern equatorial Pacific. Geofis. int., *2:*77–92.
1963. Sea level variations at Panama. J. geophys. Res., *68:*5701–5710.

ROGERS, WOODES
1712. A cruising voyage round the world, . . . London.

ROSENBLATT, RICHARD H.
1963. Some aspects of speciation in marine shore fishes. *In* Speciation in the sea. Systematics Assoc., Publ., (5):171–180, 2 figs., 1 table.

ROSENBLATT, RICHARD H. AND WAYNE J. BALDWIN
1958. A review of the eastern Pacific sharks of the genus *Carcharhinus* with a redescription of *C. malpeloensis* (Fowler) and California records of *C. remotus* (Duméril). Calif. Fish Game, *44*(2):137–159, 19 figs., 2 tables.

ROSENBLATT, RICHARD H. AND BOYD W. WALKER
1963. The marine shore-fishes of the Galápagos Islands. *In* Galápagos Islands, a unique area for scientific investigation. Occas. Pap. Calif. Acad. Sci., (44):97–106.

RUBINOFF, IRA
1963. Morphological comparisons of shore fishes separated by the Isthmus of Panama. Ph.D. thesis, Harvard Univ., Cambridge, Mass., 199 pp., 6 pls., 12 figs., 25 tables.

RUBIO Y MUÑOZ-BOCANEGRA, ANGEL
1949. Notas sobre geología de Panamá. Imprenta Nacional, Panamá, 118 pp.

RÜE, E. AUBERT DE LA
1933. Observations géologiques sur les valleés du Yuramangui et du Naya. Revue Géogr. phys. Géol. dyn., *6:*191–201.
1934. Une expédition au Yurumangui et au Naya, fleuves de la cordillére occidentale des Andes de Colombia. Géographie, *41:*17–33; 117–128, 2 maps, 8 figs.

RUIZ, S. R.
   1966. Informe sobre la pesca en Panamá. Ministerio de Agricultura, Comer-
            cio, e Industrias, Departamento de Pesca, Panamá, R. P.: 1–8, 9 tables,
            1 map (mimeo).

SACHET, M. H.
   1962a. Geography and land ecology of Clipperton Island. Atoll Res. Bull.,
            (86) :115 pp.
   1962b. Monographie physique et biologique de l'Île Clipperton. Annls. Inst.
            océanogr., Monaco, *40*(1) :107 pp.

SAENZ, WILLIAM
   1962. Fishery progress in Colombia. Proc. Gulf Caribb. Fish. Inst., (15) :140–
            144.

SAUER, WALTER
   1950. Contribuciones para el conocimiento del cuaternario en el Ecuador.
            Pt. 1. Imp. de la Universidad Central, Quito, 40 pp.

SAVILLE, MARSHALL H.
   1903. Archeological researches on the coast of Ecuador. Verh. XVI int.
            Amerikanisten-Kongress, Vienna: 331–345.
   1907. The antiquities of Manabí, Ecuador. A preliminary report. Contribu-
            tions to South America archeology. The George G. Heye Expedition.
   1910. The antiquities of Manabí, Ecuador. Irving Press, N.Y., 284 pp.
   1917. A letter of Pedro de Alvarado relating to his expedition to Ecuador.
            Museum of the American Indians. Heye Foundation, New York, 6 pp.

SCHAEFER, M. B.
   1955. Scientific investigations of the tropical tuna resources of the eastern
            Pacific. *In* Papers of the International Technical Conference on the
            Conservation of the Living Resources of the Sea. United Nations, N.Y.,
            A/Cong., 10L. II:194–221, Figs. 28–37.

SCHAEFER, M.B., M.Y. BISHOP, AND G.V. HOWARD
   1958. Algunos aspectos del afloramiento en el Golfo de Panamá. Bull. Inter-
            Amer. trop. Tuna Comm., *3*(2) :112–152.

SCHMIDT, J.
   1925. On the contents of oxygen in the ocean on both sides of Panama.
            Science, N.Y., *61*(1588) :592–593.

SCHMITT, WALDO L. AND L. P. SCHULTZ
   1940. List of fishes taken on the Presidential Cruise of 1938. Smithson.
            misc. Collns., *98*(25).

SCHOTT, GERHARD
   1932. The Humboldt Current in relation to land and sea conditions on the
            Peruvian coast. Geography, *17*(96), pt. 2, June.

SCHOTTELIUS, J. W.
   1941. Estado actual de la arqueología colombiana. Educación (Bogotá),
            1:9–24.

SCHWEIGGER, E.
1958. Upwelling along the coast of Peru. J. oceanogr. Soc. Japan, *14*(3):87–91.

SCRUGGS, W. L.
1901. The Colombian and Venezuelan republics with notes on other parts of Central and South America. Little, Brown and Co., Boston, 350 pp.

SEALE, ALVIN
1940. Report on fishes from Allan Hancock expeditions in the California Academy of Sciences. Rep. Allan Hancock Pacif. Exped., 1932–38, *9:*1–46, Pls. 1–5.

SHEPPARD, GEORGE
1927a. Further observations on the clay pebble bed on Ancón, Ecuador. Geol. Mag., *54*.
1927b. Geological observations on Isla de la Plata, Ecuador, South America. Am. J. Sci., *13:*480–486.
1928a. Notes on the Miocene of Ecuador. Bull. Am. Ass. Petr. Geol., *12*(6).
1928b. Quert deposits in Ecuador. Geol. Mag., *65*.
1928c. The geology of Ancón Point, Ecuador. J. Geol., *36*(2).
1929. Marine planation in Ecuador. Pan-Am. Geol., Sept., 1929.
1930a. The geology of South-west Ecuador. Bull. Inst. Am. Petr. Geol., *14*, March.
1930b. Notes on the climate and physiography of southwestern Ecuador. Geogrl. Rev., N.Y., *20:*445–453, 8 figs.
1930c. The igneous rocks of southwestern Ecuador. J. Geol., *38*(4).
1931a. Abnormal occurrence on the littoral of the Santa Elena Peninsula, Ecuador. Geogrl. Rev., N.Y., *21*(3):490.
1931b. Bibliografía de la geología del Ecuador. An. Univ. cent. Ecuador, *46*(276).
1932. The salt industry in Ecuador. Geogr. Rev., *22*, July.
1933. The rainy season of 1932 in southwestern Ecuador. Geogr. Rev., *23:*210–216, 5 figs.
1937. The geology of southwestern Ecuador. Thomas Murby and Co., London, 275 pp.

SIEVERS, WILHELM
1914. Reise in Peru und Ecuador ausgeführt 1909. Wiss. Veröffentl. Gesellsch. Erdkunde Leipzig. Bd. 8. Duncker & Humboldt, München and Leipzig.
1931. Geografía de Ecuador, Colombia, y Venezuela. Transl. from German by Carlos de Solas. Biblioteca de Iniciación Cultural, Barcelona, 207 pp.

SINCLAIR, J. H. AND C. P. BERKEY
1923. Cherts and igneous rocks of the Santa Elena Peninsula, Ecuador. Bull. Am. Inst. Min. metall. Engrs., (1270 M).

SIVERTSEN, E.
1933. The Norwegian Zoological expedition to the Galápagos Islands 1925, conducted by Alf Wolleback. VII. Littoral Crustacea Decapoda from the Galápagos Islands. Meddr. zool. Mus., Oslo, (38):1–38.

SMAYDA, T. J.
  1963. A quantitative analysis of the phytoplankton of the Gulf of Panama. I. Results of the regional phytoplankton surveys during July and November 1957 and March 1958. Bull. Inter-Amer. trop. Tuna Comm., 7(3):191–277.

SMITH, ROBERT E., DONALD P. DE SYLVA, AND RICHARD A. LIVELLARA
  1964. Modification and operation of the Gulf I-A high-speed plankton sampler. Chesapeake Sci., 5(1–2):72–76.

SNODGRASS, ROBERT E. AND EDMUND HELLER
  1903. Papers from the Hopkins-Stanford Galápagos Expedition, 1898–1899. XV. New fishes. Proc. Wash. Acad. Sci., 5:189–229, 19 pls.
  1905. Papers from the Hopkins-Stanford Galápagos Expedition, 1898–1899. XVII. Shore fishes of the Revillagigedo, Clipperton, Cocos, and Galápagos islands. Proc. Wash. Acad. Sci., 6:333–427.
  1906. Shore fishes of the Revillagigedo, Clipperton, Cocos, and Galápagos islands. *In* Cocos and Galápagos islands, 1898–1899. Washington, 95 pp.

STARKS, EDWIN C.
  1906. On a collection of fishes made by P. O. Simons in Ecuador and Peru. Proc. U. S. natn. Mus., 1906, 30(1468):761–800, 2 pls., 10 figs.

STEINDACHNER, FRANZ
  1875a. Ichthyologische Beiträge. II. Sber. Akad. Wiss. Wien, 1875, 71:443–480.
  1875b. Zur Fischfauna von Panamá. Ichth. Beiträge. Sber. Akad. Wiss. Wien, 72(4):65 pp.
  1876. Ichthyologische Beiträge. III. Sber. Akad. Wiss. Wien, 72:29–96, Pls. 1–8.
  1876b. Zur Fischfauna von Panamá, Acapulco und Mazatlán. Über einige neue Fischarten, insbesondere Characinen und Siluroiden aus dem Amazonenstrome. Ichth. Beiträge. Sber. Akad. Wiss. Wien, 74(5):192 pp.
  1876c. Ichthyologische Beiträge. IV. Sber. Akad. Wiss. Wien, 1875, (1876), 72(1):551–616, 13 pls.
  1877. Ichthyologische Beiträge. V. Sber. Akad. Wiss. Wien, 1876 (1877), 74(1):49–240, 15 pls.
  1878. Ichthyologische Beiträge. VI. Sber. Akad. Wiss. Wien, 77:379–392, Pls. 1–3.
  1880. Ichthyologische Beiträge. VIII. Sber. Akad. Wiss. Wien, 80:119–191, Pls. 1–3.
  1900. Vorläufiger Bericht über einige von ihrer königlichen Hoheit Frau Prinzessin Therese von Bayern während einer Reise nach Südamerika gesammelten neuen Fischarten. Anz. Akad. Wiss. Wien, 37:206–2–8.
  1902. Herpetologische und ichthyologische Ergebnisse einer Reise nach Südamerika, mit einer Einleitung von Therese Prinzessin von Bayern. Denkschr. Akad. Wiss. Wien, 72:89–148, 5 pls., 2 figs.

STÜBEL, ALPHONS (MORITZ)
  1906. Die Vulkanberge von Colombia. Geologisch-topographisch Aufgenommen und Beschreibung nach dessen Tode ergänzt und herausgegeben von Theodor Wolf. Wilhelm Baenach, Dresden, 153 pp.

SUND, P. A. AND J. A. RENNER
1959. Los quetognatos de la expedición EASTROPIC, con apuntes sobre su posible valor como indicadores de las condiciones hidrográficas. Bull. Inter-Amer. trop. Tuna Comm., *3*(9):423–538.

SVENSON, HENRY K.
1936. Vegetation of the coast of Ecuador and Peru and its relation to the Galápagos Islands. Contr. Brooklyn bot. Gdn., (104).

TARGINONI-TOZZETTI, ADOLFO
1877. Zoologia del viaggio intorno al globo della R. Pirocorvetta MAGENTA durante gli anni 1865–1868. Publ. Studi sup. prat. Firenze, 1877, (1):xxiv + 257 pp., 13 pls.

TERRY, ROBERT A.
1941. Notes on submarine valleys off the Panama coast. Geogrl. Rev., N.Y., *31*(5):377–384.
1956. A geological reconnaissance of Panama. Occ. Pap. Calif. Acad. Sci., (23):1–91.

THAYER, JOHN E.
1905. The vertebrata of Gorgona Island, Colombia. Aves. Bull. Mus. comp. Zool. Harv., *46*:91–98.

THAYER, JOHN E. AND OUTRAM BANGS
1905. The mammals and birds of the Pearl Islands, Bay of Panama. Bull. Mus. comp. Zool. Harv., *46*(8):137–160.

TORRES MARIÑO, RAFAEL
1938. Climatología colombiana, geografía fínca. Editorial Lumen. S. A., Bogotá, 110 pp.

TORTONESE, ENRICO
1939. Su alcuni Plagiostomi e Teleostei raccolti dal Dott. E. Festa nell' l'America centrale e meridionale. Boll. Musei Zool. Anat. comp. R. Univ. Torino, *47*(89):43–56.
1947–1949. Materiale per lo studio sistematico e zoogeografico dei pesci delle coste occidentali del Sud America. Revta. chil. Hist. nat., *51–53*:83–118.

TROLL, C.
1931. Ecuador. *In* Handbuch der geographische Wissenschaften Südamerika, *13*:392–411.

ULLOA, DON ANTONIO DE
1771. Historische reisbeschryving van geheel zuid-America; . . . Jacobus van Karnebeek, 's Gravenhage, *2*:406 pp.

UNESCO/UN
1961. Contribución a la bibliografía latinoamericana sobre biología marina (1955–1960). 106 pp.

URIBE, JOAQUIN ANTONIO
1935. El niño naturalista. Imprenta Departamental, Medellín, 436 pp.
1936. Cuadros de la naturaleza. Editorial Minerva, Bogotá, 168 pp. 3rd Ed.

U. S. NAVAL OCEANOGRAPHIC OFFICE
1965.   Sailing directions for the west coasts of Mexico and Central America.
        Publ. U.S. Naval oceanogr. Office, 26.

VANDERBILT, WILLIAM K.
1927.   To Galápagos on the ARA, 1926; the events of a pleasure-cruise to the
        Galápagos Islands and a classification of a few rare aquatic findings,
        including two specimens of a new species of shark never caught before
        and here described for the first time. Privately printed, W. F. Rudge, Mt.
        Vernon, N.Y., 136 pp., 30 col. pls.

VAUGHAN, T. W.
1919.   Contributions to the geology and paleontology of the Canal Zone,
        Panama, and geologically related areas in Central America and the
        West Indies. Bull. U.S. natn. Mus., (103):xviii + 612 pp.

VEGAS, M.
1963.   Contribution to the knowledge of the shallow-water zone of the Peruvian
        Coast.V. An. cient., *1*(2):174–193.

VELASCO, JUAN DE
1841.   Historia del Reino de Quito en la América meridional. *2*, pt. 2, Imprenta
        de Gobierna, Juan Campuyane, Quito, 210 pp.
1958.   Historiadores ecuatorianos. Biblioteca del Estudiante, (21). Ministerio
        de Educación Pública, Talleres Gráficos de Educación, Quito, 59 pp.

VON HAGEN, VICTOR WOLFGANG
1940.   Ecuador the unknown. Oxford Univ. Press, N.Y.
1945.   South America called them—Explorations of the great naturalists: La
        Condamine, Humboldt, Darwin. Spruce, N.Y.
1949.   Ecuador and the Galápagos Islands. Univ. Oklahoma Press, Norman,
        ix + 290 pp.
1960.   The ancient sun kingdoms of the Americas. World Publ. Co., Cleveland
        and New York, 617 pp.

VOSS, GILBERT L.
1967.   Bioenvironmental and radiological safety feasibility studies, Atlantic-
        Pacific Interoceanic Canal. Prepared for Battelle Memorial Inst., Atomic
        Energy Commission prime contract No. AT(26–1)–171. Institute of
        Marine Sciences, University of Miami, 143 pp., processed.

WAFER, LIONEL
1699.   A new voyage and description of the isthmus of America. Giving an
        account of the author's abode there, the form and make of the country
        . . . . J. Knapton, London, 224 pp.
1706.   Les voyages de Lionel Wafer. Claude Cellier, Paris, 398 pp.

WAGNER, MORITZ F.
1861.   Beiträge zu einer physischgeographisch Skizze des Isthmus von Panamá.
        Petermann's geogr. Mitt., Erganzungsheft, 125 pp.
1865a. Neue Gattungen und Arten von Fischen aus Central-Amerika, beschrie-
        ben von Kner und Steindachner. Munich, 6 pls.
1865b. Über die hydrographischer Verhältnisse und das Vorkommen der

Süsswasserfische in den Staaten Panamá und Ecuador. Abh. bayer. Akad. Wiss. Math.-phys. Kl., 1864, *10*:63–113.

WALFORD, LIONEL A.
1937. Marine game fishes of the Pacific coast from Alaska to the Equator. Univ. Calif. Press, Berkeley, 205 pp., 69 pls., 1 map.

WALKER, ALEXANDER
1822a. Colombia: being a geographical, statistical, agricultural . . . account of that country, adapted for the general reader. Baldwin, London, unpaged.
1822b. Colombia: siendo una relación geográfica topográfica, agricultural, commercial, política, & c. de aquel pais, adaptada para todo lector en general y para el comerciante y colono en particular. Baldwin, London, 2 vols.

WAVRIN, MARQUIS DE
1936. Apport aux connaissances de la civilization dite "de San Agustin" et a l'archéologie de sud de la Colombie. Bull. Soc. Americanistes belg., *21*:107–134.

WEST, ROBERT COOPER
1951. The Pacific lowlands of Colombia; a negroid area of the American tropics. Louisiana State Univ. Press, Baton Rouge, xiv + 278 pp.

WHITAKER, ARTHUR P.
1948. The United States and South America, the northern republics. Harvard Univ. Press, Cambridge, xix + 278 pp.

WHITE, E. I.
1927. On a fossil cyprinodont from Ecuador. Ann. Mag. nat. Hist., 1927: 269.

WILLIAMSON, EDWARD B.
1918. A collecting trip to Colombia, South America. Univ. Mich. Mus. Zool., misc. Publ., (3):5–24.

WILSON, CHARLES E.
1916. Some marine fishes from Colombia and Ecuador. Ann. Carneg. Mus., *10*(1–2):57–70.

WILSON, J. S.
1886. Geological notes on the Pacific coast of Ecuador, and on some evidence of the antiquity of man in that region. Q. Jl. geol. Soc. Lond., *22*(1): 567–570.

WILTON, JANSON A.
1949. La pesca en la costa del Pacífico. Calif.

WOLF, TEODORO
1869–1876. Relación de un viaje geognóstico por la provincia del Guayos. *In* M. Pissis *et al.*, Géologie de Sud-Amérique.
1879a. Memoria sobre la geografía de la provincia de Esmeraldas. Guayaquil.
1879b. Viajes científicos por la República del Ecuador. Bull. Mus. comp. Zool. Harv., *28*(5):151–285.
1892. Geografía y geología del Ecuador publicada por orden del supremo Gobierno de la República. F. A. Brockhaus, Leipzig, xxi + 671 pp.

WOOSTER, WARREN S.
   1959. Oceanographic observations in the Panama Bight, ASKOY Expedition,
          1941. Bull. Am. Mus. nat. Hist., *118*(3):113–152.

WOOSTER, WARREN S. AND T. CROMWELL
   1958. An oceanographic description of the eastern tropical Pacific. Bull.
          Scripps Instn. Oceanogr., *7:*169–282.

WYRTKI, KLAUS
   1964. Upwelling in the Costa Rica Dome. Fishery Bull. Fish Wildl. Serv., U.S.
          63(2):355–372
   1965. Corrientes superficiales del océano Pacífico oriental tropical. Bull. Inter-
          Amer. trop. Tuna Comm., *9*(5):295–424.
   1965. The thermal structure of the eastern Pacific dome. Dt. hydrogr. Z.,
          Ergänzungheft, (A) *6:*1–84.